Enlightenment Travel and British Identities

Thomas Pennant by Thomas Gainsborough (1776)
National Museum Wales

Enlightenment Travel and British Identities

Thomas Pennant's Tours in Scotland and Wales

Edited by Mary-Ann Constantine and
Nigel Leask

A

ANTHEM PRESS

Anthem Press
An imprint of Wimbledon Publishing Company
www.anthempress.com

This edition first published in UK and USA 2019
by ANTHEM PRESS
75–76 Blackfriars Road, London SE1 8HA, UK
or PO Box 9779, London SW19 7ZG, UK
and
244 Madison Ave #116, New York, NY 10016, USA

First published in the UK and USA by Anthem Press 2017

British Library Cataloguing-in-Publication Data
A catalogue record for this book is available from the British Library.

ISBN-13: 978-1-78527-177-9 (Pbk)
ISBN-10: 1-78527-177-6 (Pbk)

This title is also available as an e-book.

For JB, whether he likes it or not

He's a *Whig*, sir, a *sad dog* [...] But he's the best traveller I ever read; he observes more things than anyone else does.

Samuel Johnson

CONTENTS

FIGURES

CONTRIBUTORS

C. Stephen Briggs is an independent researcher and former head of archaeology at the Royal Commission on the Ancient and Historical Monuments of Wales. He has a long-standing interest in the manuscript and printed antiquarian and early tourist literature of Wales; has written widely on the history of archaeology, on British-Irish prehistory and on garden archaeology.

Mary-Ann Constantine is Reader at the University of Wales Centre for Advanced Welsh and Celtic Studies. She works on Romantic-period Wales and Brittany, and is currently writing a book on travellers to Wales 1760–1820. She is Principal Investigator on the AHRC-funded 'Curious Travellers' project.

Elizabeth Edwards is a Research Fellow at the University of Wales Centre for Advanced Welsh and Celtic Studies. She is currently working on the AHRC-funded 'Curious Travellers' project, editing Romantic-period tours of Wales.

R. Paul Evans is Deputy Headmaster and teacher of history at Denbigh High School. He has written widely on Thomas Pennant and his correspondents and is currently writing a biography of Thomas Pennant.

Tom Furniss is Senior Lecturer in English Studies at the University of Strathclyde. His research interests include eighteenth-century travel writing about Scotland, the textual representation of place in Romantic Scotland, and geological controversy in and about Scotland between 1750 and 1820.

Jane Hawkes is Professor in the History of Art at the University of York, with a special interest in the iconography of Anglo-Saxon sculpture. She is currently working on the historiography of Insular sculpture, with particular reference to later antiquarian and medievalist interpretations.

Ailsa Hutton has recently received her PhD from the University of Glasgow. Her research interests include antiquarianism (with a particular emphasis on forms of visual documentation), the making of topographical art and knowledge and travel-writing in Scotland during the long eighteenth century.

Dafydd Johnston is the Director of the University of Wales Centre for Advanced Welsh and Celtic Studies. He has published extensively on Welsh literature of all periods, and has a particular interest in medieval Welsh poetry.

Caroline Kerkham is an independent researcher with expertise in tourist literature relating to Wales and garden history. She has an especial interest in the picturesque and archaeological landscape of the eighteenth-century Hafod estate in Ceredigion.

Allison Ksiazkiewicz is an artist and independent scholar working on the entwined nature of the history of science and art. She is currently writing a book on British geology, antiquarianism, and the visualization of geolandscapes during the eighteenth and nineteenth centuries.

Nigel Leask is the Regius Chair of English Language and Literature at the School of Critical Studies in the University of Glasgow. He has written widely on Scottish literature, travel and empire in the eighteenth and nineteenth centuries. He is currently co-investigator on the AHRC-funded 'Curious Travellers' project and is writing a book on travellers to Scotland.

Helen McCormack is a lecturer in the Forum for Critical Inquiry at the Glasgow School of Art. She teaches courses in the history and theory of art and design from the eighteenth century to the present and is currently writing a book on the cultural networks of the eighteenth-century anatomist Dr William Hunter.

PREFACE

This volume would not have come about without the support of the British Academy, whose Small Grants scheme allowed us to run two experimental cross-disciplinary workshops focusing on the Welsh and Scottish *Tours* of Thomas Pennant. The response to these two events, held in Glasgow and Aberystwyth in 2013, was genuinely inspiring: the act of assembling archaeologists, local historians, naturalists, literary critics, art historians and digital cartographers for the workshops brought out the extraordinary variety of Pennant's travel writing. This volume of essays, selected from a range of papers on offer, captures something, though by no means all, of the multivocal nature of these rich and relatively neglected texts. We hope it will go some way towards stimulating interest not only in the life and works of Pennant himself, but in the tour as a literary genre – intellectually rich, often intriguing, sometimes frustrating, but always full of surprises.

Our thanks are due to all who took part in these workshops, whether as speakers or as audience; to the Thomas Pennant Society for their enthusiastic support; to the National Library of Wales for the wonderful venue and for the generous participation of their staff; to the University of Glasgow Library Special Collections; and to the Hunterian Museum in Glasgow. Thanks too are due to Dr Alexander Deans and Dr Angharad Elias for their help in organizing the events, and to Dr Ailsa Hutton for her help with the illustrations. We are also much indebted to Dr Gwen Gruffudd for her painstaking copy-editing of these essays, and to the staff at Anthem Press for their expertise and support. Work on Pennant and the domestic tour continues to develop thanks to the award of a substantial grant from the Arts & Humanities Research Council (AHRC), which is funding a four-year project entitled 'Curious Travellers: Thomas Pennant and the Welsh and Scottish Tour, 1760–1820'. We are extremely grateful to them for having seen the potential in this subject.

The book is dedicated to John Barrell, teacher, colleague and friend, whose brilliant work on landscape, politics and topography underpins so much of what we do: we hope that, in spite of everything, and following in the steps of Edward Pugh of Ruthin, he will come to appreciate the solid virtues of 'that faithful writer, the late Mr Pennant Esq.'. The editors would also like to thank their families for their tolerance (and in some cases blissful ignorance) of Pennant and co., and to warn them that the Walks are not over yet …

Mary-Ann Constantine, Aberystwyth
Nigel Leask, Glasgow

ACKNOWLEDGEMENTS

National Museum Wales: Frontispiece
The Society of Antiquaries of London: Figure 2.1a
National Library of Wales: Figures 2.3, 10.4 and 11.1
Jane Hawkes: Figure 4.2a
British Library Board: Figures 4.5 and 9.1
University of Glasgow Library: Figures 3.1, 6.1, 8.1 and 8.2
Geological Society of London: Figures 9.2 and 9.4
Syndics of Cambridge University Library: Figure 9.3
Hunterian Museum and Art Gallery, Glasgow: Figures 10.1 and 10.3
British Museum: Figure 10.5

ABBREVIATIONS

BL	British Library
DWB Online	'Dictionary of Welsh Biography' at http://yba.llgc.org.uk/en/index.html
NLS	National Library of Scotland
NLW	National Library of Wales
ODNB Online	'Oxford Dictionary of National Biography' at http://www.oxforddnb.com
RCAHMW	Royal Commission on the Ancient and Historical Monuments of Wales
SAL	The Society of Antiquaries of London
WCRO	Warwick County Record Office

INTRODUCTION: THOMAS PENNANT, CURIOUS TRAVELLER

Mary-Ann Constantine and Nigel Leask

I beg to be considered not as a Topographer but as a curious traveller willing to collect all that a traveller may be supposed to do in his voyage: I am the first that attempted travels at home, therefore earnestly wish for accuracy.

– Thomas Pennant, 1773[1]

The essays collected in this volume explore the crowded, multifaceted world of the Welsh naturalist, antiquarian and traveller Thomas Pennant (1726–1798). The tribute is long overdue: despite Pennant's indisputable contribution to eighteenth-century intellectual life and his subsequent influence on writers of the following century, no previous volume has ever been exclusively devoted to him.[2] Born into a Welsh gentry family from Downing in Flintshire, Pennant was educated at Wrexham Grammar School and Oxford, and in the 1750s and 1760s toured Europe, Cornwall, Ireland and Wales in search of mineral and ornithological specimens. At the time of his Scottish tours he was thus already known as a naturalist and author of *British Zoology* (1761–66) and *Synopsis of Quadrupeds* (1771). From his estate at Downing, where he lived the life of an 'improving' landlord until his death, Pennant established a national and international correspondence network, which included Joseph Banks, Richard Gough, Gilbert White of Selborne, Simon Pallas, the Dutch naturalist Gronovius, and the Swedish botanist Carl Linnaeus. Part of the reason for his comparative neglect, one suspects, is the very scattered nature of his archive, with letters (so often the key to understanding the composition of the tours) to be found in libraries, archives and private collections the length and breadth of Britain and beyond. But since Pennant was also a passionate antiquarian and a competent historian, with a keen interest in art, agriculture and industrial experimentation, it may also be the case that his rather overwhelming interdisciplinarity has deterred individual scholars from getting to grips with his life and works: though frequently cited as witness or authority in other studies, especially on the natural and social history of Scotland and Wales, his texts have rarely been addressed in their own right. One obvious response to the dilemma of Pennant's 'plurality', then, was to bring together scholars from a range of disciplines to focus on particular texts. The contributions to this volume, though far from exhausting the interpretive possibilities of his work, demonstrate just how fruitful that multiple approach can be.

'Travels at *home*': Pennant's British Identities

Pennant is now best remembered as a travel writer: his pioneering and widely read *Tours* in Scotland and Wales inspired hundreds of subsequent travellers, and influenced the emergence of a new 'British' identity which would take a more positive view of its Celtic peripheries in the century following Anglo–Scottish Union in 1707, and the turmoil of the Jacobite uprisings in 1715, 1719 and 1745. That identity, of course, was far from homogenous or consistent, and it is deeply interesting to read Pennant in the light of the work done on 'Britishness' by critics such as Linda Colley, Colin Kidd and Murray Pittock. Paul Smethurst writes of 'the spirit of suture' that prompted Pennant, like his precursor Daniel Defoe (albeit, perhaps, with a more open mind) to 'trace paths back and forth across the borders of Wales, England and Scotland'.[3] There is no doubt that Pennant thought of himself as a good 'Briton' in the Hanoverian, unionist mould – one who, in his writing on Scotland, had 'laboured earnestly to conciliate the affections of the two nations'.[4] Yet, as many of the following essays demonstrate, the political intentions (and the unintended side effects) of the domestic tour are often far from synchronized. Though the writings of Pennant and his followers appear to bring the edges and regions of Britain closer to the centre they were also developing a complex concatenation of national histories, both at a 'four nations' level, and at a larger British one, which effectively competed with the official historiography of the Enlightenment.[5] Tour writing uncovered the 'British' past in the landscape, recovering it through sites (cromlechs, churches, castles, ruins), through popular culture (language, costume, song and 'national character'), and also through the collection and study of material objects (archaeological finds). The narratives of that past, however, were far from univocal. Nor were they automatically linear, and simply committed to a progressive development from primitive pastoralism to commercial modernity, as theorized by Scottish 'stadial theorists' Adam Smith and Adam Ferguson. Fredrik Albritton Jonsson has recently drawn attention to the 'hybrid modernity' described by Pennant's acquaintance and correspondent, the Reverend John Walker in his 'Reports on the Hebrides' compiled in the 1760s: here, says Jonsson, 'the crooked spade [i.e. the traditional Gaelic hand plough], the potato, and the projected linen manufacture – mixed and jumbled the stages of conjectural history, challenging the notion of a necessary linear progression between primitive agriculture and modern manufacture'.[6] Thus, though on the one hand Highland tours and their Welsh equivalents 'offered a time machine for adherents of stadial history',[7] the story of history-as-progress was often caught up in the conflicting discourses of antiquarianism, civic patriotism and improvement.

'Travels at home', then, cannot be an innocent concept, when 'home', simultaneously perceived through past and present, is a place of multiple, sometimes directly conflicting, identities. Looking back on late eighteenth-century British consolidation with the hindsight provided by our own devolutionary era prompts us to consider the extent to which native histories of Wales and Scotland had the power to disrupt a larger narrative of uniform British identity, as Ina Ferris has suggested they did in Ireland.[8] The presence of alternative national histories, involving issues of language (notably the translation and interpretation of Welsh and Gaelic sources) became especially pressing during the decades of revolution and war. As a member of Flintshire's anglicized gentry class, Pennant's sense of his ethnicity fluctuated between unselfconscious identification with England (in a stroppy letter to George Paton, he remonstrated 'we English love conciseness'), and a deep-felt Welshness, typically expressed in the confident opening lines of the *Tour in Wales* (cited in Chapter 3) or in his admission that he is 'ancient Britain' [*sic*] enough to regret that Ben Nevis was now deemed to be higher than Snowdon.[9] The use of the words 'us' and 'we' in Pennant's Scottish and Welsh tours is always worth watching, since Celtic *filiation* can sometimes trump – or at least ruffle – British *affiliation*, permitting the emergence of a 'transperipheral' vision that competes with the improving metropolitan perspective that elsewhere dominates his narrative.[10] This dual-voicing within the text proved an intriguing resource for national-minded Welsh and Scottish writers in the Romantic period, as several essays in the present collection demonstrate – the revival of Owain Glyndŵr as a Welsh national hero being a particularly striking example.

Pennant's transperipheral perspective may go some way to explaining the largely positive reception (on both sides of the border) of his account of Scotland, which had represented a minefield for English tourists and travel writers in the decades before his 1769 tour, and would again in 1775, when Samuel Johnson's *Journey to the Western Islands of Scotland* elicited furious reactions from the Scottish public. Given the emotive treatment of the subject in later tours and official reports on the state of the Highlands, one might note in particular Pennant's remarkably unfussed attitude to the Gaelic language, which neither repels nor overly excites him. As a largely English-speaking Welshman used to living with two languages around him, Pennant (who took an interpreter on his 1772 tour) treats the use of Gaelic as wholly natural. When he draws attention to the language, it is usually (as so often with the Welsh tours) in the context of topography or natural history – in native names for places or plants, carefully transliterated in his text with the expert assistance of the Gaelic scholar Reverend John Stuart of Killin. There is, however, one intriguing episode where a social, and potentially political, linguistic situation arises:

> At seven in the morning, take a six-oared boat, at the east end of *Loch-maree*: keep on the north shore beneath steep rocks, mostly filled with pines waving over our heads. Observe on the shore a young man of good appearance, hailing the boat in the *Erse* language. I demanded what he wanted: was informed, a place in the boat. As it was entirely filled, I was obliged to refuse his request. He follows us for two miles, through every difficulty, and by his voice and gestures threatened revenge. At length a rower thought fit to acquaint us, that he was the owner of the boat, and only wanted admission in lieu of one of them. The boat was taken ashore, and the master taken in with proper apologies and attempts to sooth him for his hard treatment. Instead of insulting us with abuse, as a *Charon* of *South Britain* would have done, he instantly composed himself, and told us through an interpreter, that he felt great pride in finding that his conduct gained any degree of approbation.[11]

The passage starts with the familiar image of the alien and incomprehensible native jumping up and down with rage on the shoreline. It ends, through the mediation of the interpreter, with extraordinary civility on both sides. Both Pennant and the ferryman are expecting to receive an earful of abuse; touchingly, both are relieved and surprised to find the other side so courteously apologetic for their own conduct. If this is, consciously or otherwise, a little parable of peaceable Union, it is nevertheless one in which language difference is not figured as inherently divisive.

Worlds of Text and Image: Travel Writing and Enlightenment Networks

Given the credible claim that by 1800 travel writing was the most widely read division of literature after novels and romances,[12] the long-standing critical neglect of this most important branch of the 'literature of fact' is remarkable, although a flood of recent publications, and two new journals devoted to the subject, are swiftly changing the field.[13] The eighteenth century was the second great European age of travel and exploration, when the humanistic paradigms established a century earlier in the aristocratic Grand Tour to Italy were vastly and dramatically extended. Enlightenment travel is epitomized by the oceanic and intra-continental explorations of Cook, Banks, Bougainville, Humboldt and others, who transformed what Mary Louise Pratt termed 'European planetary consciousness' by means of rigorous practices of surveying, collection and empirical study in far-flung places; the links between Enlightenment natural history and the concept of stadial 'human' history have also received critical attention.[14] Yet, despite Pat Rogers's acknowledgement that the immediate context of Samuel Johnson's 'transit of the Caledonian hemisphere' in 1773 was public excitement about Cook's Pacific voyages, the rise of the domestic tour in the decades after 1760 has tended to be viewed in detachment from this global picture.[15] Although there is a healthy secondary literature on the domestic tour in England,[16] studies of tours of Wales and Scotland have, with few exceptions, focused on the socio-economic aspects of travellers' accounts, or read them in the context of the history of leisure tourism.[17] Apart from critical work on Johnson and Boswell, or Wordsworth and Coleridge, and some excellent study of the picturesque, relatively little scholarship has addressed the literary, scientific, antiquarian and political concerns of the domestic tour in these decades.[18]

Pennant himself linked domestic travel with its more exotic counterpart when he claimed in his *Literary Life* that, prior to his tour, Scotland had been 'almost as little known to its southern brethren as *Kamtschatka*'.[19] Although in contrast to Admiralty-supported global exploration Pennant's tours were privately funded, the second (1772) Scottish tour certainly aimed to introduce a new professional rigour into the practice and representation of domestic travel, and was inspired by the contemporaneous expeditions of Cook and Joseph Banks to the South Pacific in 1768–71. In 1772 Pennant travelled with a team comprising a botanist, a Gaelic expert and a trained topographical artist. In this respect, his *Tours* are distinct from the leisurely domestic tourism that was beginning to emerge in England and Wales, and which would reach its apogee, post-Gilpin, in the 'picturesque tours' of the early Romantic period. His 1772 Scottish tour itself overlapped with Banks' and Solander's exploration of some of the Inner Hebrides en route for Iceland, and Pennant's 'Voyage to the Hebrides' (Part 1 of his *Tour in Scotland 1772*) contains one of the few published accounts of that expedition in Banks's description of the island of Staffa. The volume itself was appropriately dedicated to 'Joseph Banks Esq.', whom Pennant praised as having 'enriched yourself with the treasures of the globe, by a circumnavigation, founded on most liberal and scientific principles'.

Banks's and Cook's reliance on travelling artists like Alexander Buchan, Sydney Parkinson and William Hodges has been well studied, yet it has been seldom remarked that Pennant aspired to a similar level of visual documentation, a point touched on in several essays in the present collection.[20] Recent research by art historians, historical cartographers and literary critics has signalled new directions in the study of the relations between image and text in topographical works.[21] Moses Griffith and other artists employed – or sometimes simply invoked – by Pennant helped to create the imagined landscape of the Welsh and Scottish tours, as when he relinquishes his attempt to capture in words a dramatic view of Cadair Idris, 'so excellently expressed by the admirable pencil of my kinsman, Mr [Richard] Wilson, that I shall not attempt the description'.[22] In addition to his published plates, a more select circle of connoisseurs could enjoy the sumptuous extra-illustrated Tours, crammed with marginal watercolours of birds, plants and ancient and modern buildings and antiquities, all of which helped shape perceptions not only of present landscapes, but of an imagined British past.[23]

Although Dr Johnson famously described Pennant as 'the best traveller I ever read; he observes more things than anyone else does', James Boswell was less complimentary, complaining that 'he shews no philosophical investigation of character and manners', and 'could put together only curt frittered fragments of his own, and afterwards procured supplemental intelligence from parochial ministers, and others not the best qualified or most impartial observers'[24] – an opinion not far from Horace Walpole's similarly negative judgement, discussed in Paul Evans's essay (Chapter 1 in this volume). Boswell's was a harsh, although substantially accurate, account of Pennant's methods of composition, and recent commentators have also noted that 'encyclopaedic detail [often threatens] to overwhelm the sense of *being there*' in his published narratives.[25] In many respects physical travel (impressive as those extraordinary journeys are) was only the tip of the iceberg for Pennant. Much of his work was conducted at home in Downing, the centre of his 'learned empire', from where he corresponded with members of his intellectual network across Britain and Europe, and drew on the goodwill and expertise of neighbours like the indefatigable John Lloyd of Caerwys. Pennant's extensive correspondence is widely dispersed, and has only recently started to be gathered together and edited.[26] A window onto the networks of his age, the letters illuminate the composition, development and reception of his *Tours*, shedding light on practical issues such as the patron–artist relationship, the logistics of printing and publishing, the incorporation of earlier material (often by other writers) the commissioning and placement of illustrations, and the addition of maps. They reveal that each volume involved an impressive assemblage of supplementary material, in both textual and visual media, to add flesh to the bones of his itinerary in preparing the published account of his tours. (Over forty informants are acknowledged in the preface to the 1772 *Tour in Scotland*, and many others had doubtless contributed information.) This massive absorption of other people's material leads to some very interesting moments of shifting narrative perspective, and nicely exemplifies an acute recent characterization of travel writing as 'highly accretional in texture, "composed" in numerous ways and part of multiple conversations between exploration and authorship as complex processes of attribution and authority.'[27] It also helps explain why Pennant's publications are a bibliographer's nightmare, as texts undergo invisible extension and supplementation, and plates and appendices change from one edition to another – all this without taking into account the 'private consumption' of the Tours, where, as discussed in Chapter 6, personal marginalia, interleaved texts and images created unique, bespoke versions which directly commented on and engaged with the printed text. A simplified bibliography of Pennant's *Tours* is included as an appendix to this volume, but the process of identifying the various editions remains a work in process.

Thanks to the skills of his own trained artist, Moses Griffith, and his collaborations with others, such as Paul Sandby, Pennant was the first travel writer to produce a richly documented visual account of Scotland and Wales, and his books accordingly transformed the eighteenth-century domestic tour into a highly saleable commodity. John Bonehill has described how 'in antiquarian and natural historical circles, letters, books, maps, plans, drawings and prints as well as objects were lent and copied in a country-wide enterprise'.[28] This kind of networking – discussed further here in Hutton and Leask's essay on Robert Riddell (Chapter 6) – was a standard feature of Enlightenment knowledge production. What was perhaps distinctive about Pennant's approach was his strongly commercial orientation, and his insistence on publishing the fruits of his own researches, as well as those of his 'elèves' such as John Lightfoot and Charles Cordiner, in an illustrated format that made them attractive to a non-specialized reading public. As Fredrik Albritton Jonsson sums it up, 'Pennant's best-selling tours of Scotland repackaged the business of strategic surveying into a form of polite entertainment', and the same might be said for his Welsh tours too.[29] And yet, as innumerable footnotes in a myriad publications well into the nineteenth century testify, Pennant was, entertainment apart, also an influential intellectual authority, a source of historical and cultural as well as scientific information. His work is cited (or simply absorbed verbatim) in publications as diverse as song collections, histories, critical editions of medieval texts, contemporary poetry and novels, besides, of course, being endlessly recycled in subsequent tour narratives.

Multiple Voices: About this Volume

Pennant's interdisciplinary range makes him particularly well suited to the format of an essay collection whose various contributors span a variety of academic disciplines, from literature, history and archaeology to art history and the history of science. After Paul Evans's informative overview of Pennant's career and the intellectual significance and practical achievements of the Scottish and Welsh tours, the remaining ten essays published in this book are arranged in two sections. The first section is devoted to history, antiquities and literature; the second section focuses on natural history and the arts, moving from geology and zoology through to botany. The arrangement is intended to be schematic only, and essays in both sections reveal various thematic overlaps: the nature of literary and visual representation, for example, is a recurring presence in nearly all of the contributions, as is a concern to understand travel writing both as a form of and vehicle for scientific or antiquarian endeavour. Discussion of Pennant's legacy – literary and scientific – appears in a number of different contexts.

The introductory essay by Paul Evans draws extensively on Pennant's own correspondence, shedding new and fascinating light on the period of his earlier travels in search of specimens, particularly in Ireland (a journey about which Pennant himself is famously laconic: 'such was the conviviality of the country, that my journal proved as *maigre* as my entertainment was *gras*, so it never was a dish fit to be offered to the public').[30] Most importantly, though, Evans charts the discernible shift, or rather broadening, in the focus of Pennant's tour writing between the 1769 Scottish tour and the expedition of 1772, as his interest in the social and economic situation in Scotland, and the manner and customs of its people, push their way with increasing urgency into accounts of the country's wildlife and natural habitats. Evans also demonstrates the importance of Pennant's correspondence networks in preparations for and subsequent writing of the Scottish tours, acknowledging in particular the key role played by the energetic, Edinburgh-based, George Paton. Paton was a key figure in a controversial and long-drawn-out saga involving the commissioning and subsequent use of work on Orkney and Shetland by the naturalist and antiquarian Reverend George Low, an episode critically revisited in this volume by Stephen Briggs (Chapter 2), whose essay also sheds new light on some of Pennant's earlier activities and networks, including his membership of the Society of Antiquaries. Briggs assesses Pennant's talents as a proto-archaeologist, evaluating the value of his descriptions of prehistoric monuments and artefacts – stone circles, flint axes, metal torques – for practitioners of the discipline today. Once again, the multi-layered nature of the texts, and their absorption of material from many different sources, can prove a challenge; yet Briggs acknowledges that Pennant the antiquarian was both a sharp observer and 'a pre-eminent and systematic collector of information and master of scholarly synthesis'.

Pennant's responses to succeeding historical layers of Britain's complex past focus the work of both Mary-Ann Constantine and Jane Hawkes, in essays dealing respectively with the Tours' presentation of Roman and Anglo-Saxon or early medieval sites and artefacts (Chapters 3 and 4). Both writers find a contemporary political undertow to this antiquarian material. Constantine argues that the Roman conquest of the native Ancient Britons/Caledonians – a campaign visibly mapped in roads and strongholds, and long familiar to Pennant through the Classical texts of Caesar and Tacitus – cannot help but act as a rather unsettling counterpoint to the overriding narrative of a unified, improved and civilized 'Britannia', recently made whole again under Hanoverian rule. Constantine explores moments of tension or slippage in the text which can be read as symptoms of 'identity strain' as the narrative seems to fluctuate between 'Roman' and 'Ancient British' perspectives. Hawkes reads Pennant's descriptions of the early Christian sculptures of northern England and Scotland (including the famous Ruthwell Cross) in the context of a shift in taste resulting from the influence of the European Grand Tour, with its focus on the acquisition and emulation of Classical art and antiquities. These vernacular carvings, long problematic (and little valued) because of their 'papist' Christian associations, could increasingly be compared to 'Classical' sculptural monuments in Rome, a terminology which both neutralized and, by association, elevated them. Pennant's work, suggests Hawkes, expressed a complex nexus of ideas and attitudes surrounding the early medieval sculptures of England and Scotland in ways that would provide future writers with a framework that enabled further discussion of this hitherto largely neglected category of the monumental arts of Britain.

Pennant's engagement with a particular historical moment is the focus of Dafydd Johnston's essay (Chapter 5), which examines the substantial and diligently researched account of the life of the fifteenth-century Welsh rebel leader Owain Glyndŵr in the first volume of the *Tour in Wales*. This account would be highly influential in establishing Glyndŵr as a national (and nationalist) hero for future generations, but, as Johnston shows, Pennant is far from one-sided or univocal in his presentation – once again, the text reveals tensions between Welsh patriotism and British loyalism. Discussing the wide range of sources behind the finished text, including contemporary documents, published chronicles in several languages, Welsh-language poetry, genealogies, oral traditions, monuments and landscape features, Johnston particularly emphasizes the invaluable contributions made by a network of assistants and informants in Wales and Oxford.

Yet another corner of that industrious web of correspondents is explored by Ailsa Hutton and Nigel Leask in their detailed account of the compilation of the magnificent extra-illustrated and annotated three-volume copy of Pennant's

Scottish *Tours* once owned by Robert Riddell of Glenriddell (1755–94), a man described by Richard Gough as 'the first antiquary of his country' (Chapter 6). Their analysis of some of Riddell's 131 extra-illustrations and annotations locates him in a local, national and international network linking antiquarians such as Lord Hailes, Adam de Cardonnel and George Paton in Edinburgh, the earl of Buchan in the Borders, Thomas Pennant in Wales, Francis Grose and Richard Gough in London, and other correspondents in Denmark, France, America and Bengal. Although generally endorsing Pennant's account of Scotland, Riddell's Scottish patriotism and Jacobite sympathies generated some sharp local criticism in the margins: this richly layered, complex and beautiful book offers a fine example of how Pennant's work proliferated, developed and took on new lives in different contexts.

Reception is also the theme of the closing essay in this section (Chapter 7), in which Elizabeth Edwards tracks down a range of allusions to Pennant in poetry written from the 1790s onwards, before focusing on poems published by William Sotheby and Anna Seward, both drawing extensively on the *Tours in Wales* in their representations of Welsh landscape and history. Edwards traces the ways in which the places and spaces of north Wales were inescapably coloured by Pennant for those who travelled through them after reading his *Tours*, and discusses potential difficulties raised by the text for their non-Welsh 'outsider' readers on the tourist trail. Examining responses to Pennant from within Wales, Edwards examines Charlotte Wardle's medievalist ballad romance *St. Aelian's, or the Cursing Well* (1814), and shows how it became possible to weave aspects of Pennant's *Tours* into the emergent Welsh cultural nationalism of the early nineteenth century.

The second part of our volume moves into the study of the natural world, opening with Tom Furniss's vivid account of Pennant's responses to the geology of the Scottish Highlands (Chapter 8). Furniss situates Pennant at the cusp of two scientific world views – that of a 'static', empirical eighteenth-century mineralogy which largely avoided theoretical speculation about the formation and historical development of the earth, and a more dynamic interpretation which envisaged landscape forms as the direct result of profound geological processes and upheavals in the distant past. Contextualizing the pioneering discoveries of Pennant's 1772 tour (the most famous of which was his recognition of the basaltic nature of the island of Staffa), Furniss argues that Pennant's careful observations of geological formations helped to make the Highlands and Islands into key locations for Romantic geology. Staffa's extraordinary basalt pillars are also at the heart of the essay by Allison Ksiazkiewicz (Chapter 9), which analyses Joseph Banks's description of the place in conjunction with a number of contemporary visual representations, in order to consider the role of Enlightenment aesthetics in accounts of

natural history. Banks used architectural analogy as a technique for rational-
izing the site, describing the basalt columns through both Neoclassical and
Gothic architectural vocabularies, which emphasized different aspects of
this strange landscape. Pennant's own account of his tour through Scotland
predominantly employed a picturesque vocabulary; however (as Furniss also
argues) tensions between aesthetic and empirical visions of Nature do exist
within his descriptions.

Affiliations between the discourse of the visual arts and the project of
Enlightenment science are also explored in Helen McCormack's essay
(Chapter 10), which puts Pennant into conversation with his close contemporaries the animal painter George Stubbs and the medic William Hunter.
McCormack describes how during the 1770s Pennant's empirical, observational approach to the natural world became explicitly connected to the work
of these two men, and how the ideas they exchanged were absorbed into
the debates 'surrounding the imitation and representation of nature within
the cosmopolitan world of the Royal Academy of Arts and polite culture in
London'. In a deft comparison linking the three men's very different spheres
of activity, McCormack shows how contemporary notions of 'truth to nature'
and direct observation which underlay the rhetoric of the travel narrative (and
which Pennant also expressed through his employment of the 'accurate' pencil of Moses Griffith) can also be read in terms of an anatomical tradition of
autopsia – 'to see with one's own eyes'.

In the volume's final chapter, Caroline Kerkham examines Pennant's legacy in the changing attitudes of botanizing tourists' observations on the landscape in the course of the long nineteenth century. From the early 1770s, a
subtle shift in the travellers' observations broadened from plant-hunting to
consideration of the landscape from natural historical, topographical, antiquarian and aesthetic perspectives. Thus, just as Pennant's questionnaire,
'Queries proposed to Gentlemen in the several parts of Great Britain' offered
nineteenth-century botanizing tourists and visitors to Wales a template upon
which to build accounts of their own scientific exploration, his more inclusive
'tour' formula led to the publication of some fine descriptive landscape writing in the century that followed. Kerkham shows how the establishment of
journals, scientific societies and field clubs provided greater opportunities for
the exchange of botanical information and ideas; and tracks too an emergent
consciousness of the vulnerability of plant habitats and the rise of ecology and
conservationism.

Pennant has been credited with one of the earliest observations of an
example of such ecological fragility: in a passage from the *Tour in Wales*
describing the effects of pollution on plants growing near the copper mines
of Mynydd Parys on Anglesey, he finds that 'even the mosses and the lichens

of the rocks have perished: and nothing seems capable of resisting the fumes but the purple *Melic* grass which flourishes in abundance'.[31] Once again, a little investigation into his correspondence suggests that the credit for this observation should perhaps go rather to the Anglesey cleric and botanist Hugh Davies – yet another of Pennant's indefatigable informants and travelling companions.[32] Here again we see how a few lines of text open outwards to include other voices, teasing the curious reader with further possibilities for exploration. Indeed, for all their solid 'authority' as guidebooks, history books and sources of information for future generations, Pennant's *Tours* always seem, in the reading, to behave in oddly organic ways – they grow, they ramify, like some continually evolving conversation. They are full of indirect narratives – paths that could still be taken. This volume maps some of those paths, and tells some of those stories, but there are many, many more, still waiting to be discovered.

Notes

1 T. Pennant to G. Ashby, 18 May 1773, MS E2/22/2/f. 25 Suffolk County Record Office.
2 For a useful introductory article, see R. Paul Evans, 'Thomas Pennant (1726–1798): The Father of Cambrian Tourists', *Cylchgrawn Hanes Cymru / Welsh History Review* 13, no. 4 (1987): 395–417. Evans is currently working on a biography of Pennant, and has published a number of articles deriving from his earlier research: 'Thomas Pennant's Writings on North Wales' (unpublished MA thesis, University of Wales, 1985) and 'The Life and Work of Thomas Pennant (1726–1798)' (unpublished PhD thesis, University of Wales, 1993). Peter Bishop, *The Mountains of Snowdonia in Art: The Visualisation of Mountain Scenery from the Mid-Eighteenth Century to the Present Day* (Llanrwst, 2015), likewise draws on earlier masters and doctoral research and foregrounds Pennant's role. The Flintshire-based Thomas Pennant Society has also done much to raise the writer's profile locally.
3 Paul Smethurst, *Travel Writing and the Natural World, 1768–1840* (Houndmills: Palgrave Macmillan, 2013), 119. See Linda Colley, *Britons: Forging the Nation, 1707–1837* (London: Pimlico, 1994); Colin Kidd, *British Identities Before Nationalism: Ethnicity and Nationhood in the Atlantic World, 1600–1800* (Cambridge: Cambridge University Press, 1999); Murray Pittock, *Inventing and Resisting Britain: Cultural Identities in Britain and Ireland, 1685–1769* (Basingstoke: Macmillan, 1997).
4 Thomas Pennant, *The Literary Life of the Late Thomas Pennant, Esq., by Himself* (London, 1793), 1.
5 In 1780 Horace Walpole noted a 'jump from ornithology to antiquity' in Pennant's Scottish and Welsh Tours (see Paul Evans's essay for a detailed account.) For Pennant and historiography, see Mary-Ann Constantine, '"To Trace Thy Country's Glories to Their Source": Dangerous History in Thomas Pennant's *Tour in Wales*', in *Rethinking British Romantic History 1770–1845*, ed. Porscha Fermanis and John Regan (Oxford: Oxford University Press, 2015), 121–43.
6 Frederik Albritton Jonsson, *Enlightenment's Frontier: The Scottish Highlands and the Origins of Environmentalism* (New Haven and London: Yale University Press, 2013), 38.
7 Ibid. 50.

8 See Ina Ferris, *The Romantic National Tale and the Question of Ireland* (Cambridge: Cambridge University Press, 2004), 18–45.

9 George Low, *A Tour through the Islands of Orkney and Shetland*, with an introduction by Joseph Anderson (Kirkwall, 1879), lviii; *Tour in Scotland, 1769* (Chester, 1771), 207.

10 Compare Paul Smethurst's argument in 'Peripheral Vision, Landscape, and Nation Building in Thomas Pennant's Tours of Scotland, 1769–72', in *Travel Writing and Tourism in Britain and Ireland*, ed. Benjamin Colbert (London: Palgrave Macmillan, 2012), 13–30 (19). P. Hately Waddell, in his bizarre study *Ossian and the Clyde [...] or, Ossian Historical and Authentic* (Glasgow: Maclehose, 1875), refers to Pennant as 'the entertaining Englishman' (86); more recently, in an otherwise flawless essay, Susan Manning writes of 'the English antiquary and naturalist Thomas Pennant', in 'Antiquarianism, Balladry, and the Rehabilitation of Romance', in *The Cambridge History of English Romantic Literature*, ed. James Chandler (Cambridge: Cambridge University Press, 2009), 47.

11 Thomas Pennant, *A Tour in Scotland, and Voyage to the Hebrides, MDCCLXXII* (Chester, 1774), 380–81.

12 Charles Batten, *Pleasurable Instruction: Form and Convention in 18th Century Travel Literature* (Berkeley and London: University of California Press, 1978), 1.

13 *Studies in Travel Writing* (Routledge) and *Journeys: The International Journal of Travel and Travel Writing* (Berghahn).

14 Mary Louise Pratt, *Imperial Eyes: Travel Writing and Transculturation* (London and New York: Routledge, 1992); Nigel Leask, *Curiosity and the Aesthetics of Travel Writing, 1770–1840: 'From an Antique Land'* (Oxford: Oxford University Press, 2002); John Gascoigne, *Science in the Service of Empire: Joseph Banks, the British State, and the Uses of Science in the Age of Revolutions and Empire* (Cambridge: Cambridge University Press, 1998); Jonathan Lamb, *Saving the Self in the South Pacific, 1680–1840* (Chicago: Chicago University Press, 2001); B. Tobin, *Colonizing Nature: The Tropics in British Arts and Letters, 1660–1820* (Philadelphia: University of Pennsylvania Press, 2004); Smethurst, *Travel Writing and the Natural World*; David Miller and Peter Reill, *Visions of Empire: Voyages, Botany and Representations of Nature* (Cambridge: Cambridge University Press, 1996).

15 Pat Rogers, *Johnson and Boswell: The Transit of Caledonia* (Oxford: Clarendon Press, 1995).

16 Esther Moir, *The Discovery of Britain: The English Tourist 1540–1840* (London: Routledge & Kegan Paul, 1964); Ian Ousby, *The Englishman's England: Taste, Travel and the Rise of Tourism* (Cambridge: Cambridge University Press, 1990); Carole Fabricant, 'The Literature of Domestic Tourism and the Public Consumption of Public Property', in *The New Eighteenth Century: Theory, Politics, English Literature*, ed. Felicity Nussbaum and Laura Brown (New York: Methuen, 1987), 254–75; Nicola Watson, *The Literary Tourist* (Basingstoke: Palgrave Macmillan, 2006).

17 Exceptions are Peter Womack's excellent study *Improvement and Romance: Constructing the Myth of the Highlands* (Basingstoke: Macmillan, 1988), and John Glendening's *The High Road: Romantic Tourism, Scotland, and Literature, 1720–1820* (Houndmills: Macmillan, 1997). See also Martin Rackwitz, *Travels to Terra Incognita* (Münster: Waxmann, 2007); Alastair J. Durie, *Scotland for the Holidays: A History of Tourism in Scotland, c.1780–1939* (East Linton: Tuckwell Press, 2005); Katherine Haldane Grenier, *Tourism and Identity in Scotland, 1770–1914: Creating Caledonia* (Farnham, Surrey and Burlinton, VT: Ashgate, 2005); Betty Hagglund, *Tourists and Travellers: Women's Non-Fictional Writing about Scotland, 1770–1830* (Bristol: Channel View, 2010); Michael Freeman, 'In Search of the Picturesque in Wales 1770–1830', *The Picturesque* 70 (2009): 27–40; Liz Pitman, *Pigsties and Paradise: Lady Diarists and the Tour of Wales 1795–1860* (Llanrwst: Gwasg Carreg

Gwalch, 2009). In the introduction to his essay collection *Travel Writing and Tourism in Britain and Ireland*, Benjamin Colbert notes that 'it is the first devoted solely to the home tour' (3).

18 Malcolm Andrews, *The Search for the Picturesque: Landscape Aesthetics and Tourism in Britain, 1760–1800* (Stanford: Stanford University Press, 1989); Stephen Copley and Peter Garside, eds., *The Politics of the Picturesque: Literature, Landscape and Aesthetics since 1770* (Cambridge: Cambridge University Press, 1994); Elizabeth Bohls, *Women Travel Writers and the Language of Aesthetics, 1716–1818* (Cambridge: Cambridge University Press, 1995); Ann Bermingham, *Landscape and Ideology: The English Rural Tradition, 1740–1860* (Berkeley, California, and Los Angeles: University of California Press, 1986).

19 Pennant, *Literary Life*, 11.

20 Rudiger Joppien and Bernard Smith, *The Art of Captain Cook's Voyages* (New Haven: Yale University Press, 1985–7); Bernard Smith, *European Vision and the South Pacific* (Oxford: Oxford University Press, 1960; 2nd ed. New Haven: Yale University Press, 1985); Harriet Guest, *Empire, Barbarism, and Civilization: James Cook, William Hodges, and the Return to the Pacific* (Cambridge: Cambridge University Press, 2007).

21 John Barrell, *Edward Pugh of Ruthin 1763–1813: 'A Native Artist'* (Cardiff: University of Wales Press, 2013); Noah Heringman, *Sciences of Antiquity: Romantic Antiquarianism, Natural History and Knowledge Work* (Oxford: Oxford University Press, 2013); Sam Smiles, *Eyewitness: Artists and Visual Documentation in Britain, 1770–1830* (Aldershot: Ashgate 2000); C. W. J. Withers, *Geography, Science and National Identity: Scotland since 1520* (Cambridge: Cambridge University Press, 2001).

22 Thomas Pennant, *A Tour in Wales*, 2 vols (London, 1784), II: 98. Wilson's striking 'Llyn y Cau, Cader Idris' (*c.*1774) is held by the Tate Gallery.

23 A splendid eight-volume extra-illustrated version of the *Tour in Wales* is available online through the National Library of Wales's website: www.llgc.org.uk/digitalmirror (accessed 30 May 2015).

24 James Boswell, *Life of Dr Samuel Johnson*, ed. R. W. Chapman, rev. J. D. Fleeman, with a new introduction by Pat Rogers (Oxford: Oxford University Press, 1980), 933.

25 Smethurst, 'Peripheral Vision, Landscape, and Nation-Building', 18; Constantine, ' "To Trace Thy Country's Glories to their Source" ', 126.

26 For more information about an ongoing AHRC-funded project focused on Pennant's tours and correspondence, see www.curioustravellers.ac.uk (accessed 30 May 2015).

27 Innes M. Keighren, Charles W. J. Withers and Bill Bell, *Travels into Print: Exploration, Writing and Publishing with John Murray, 1773–1859* (Chicago and London: The University of Chicago Press, 2015), 4.

28 John Bonehill, ' "New scenes drawn by the pencil of Truth": Joseph Banks's Northern Voyage', *Journal of Historical Geography* 43 (2014): 9–27 (13).

29 Jonsson, *Enlightenment's Frontier*, 50.

30 Pennant, *Literary Life*, 2.

31 Idem, *Tour in Wales* (1784), II: 275.

32 Noted by Gareth Griffith, 'Cysylltiadau Biolegol: Hugh Davies a Thomas Pennant' (unpublished paper, Thomas Pennant Workshop, Aberystwyth, June 2013).

Chapter 1

'A ROUND JUMP FROM ORNITHOLOGY TO ANTIQUITY': THE DEVELOPMENT OF THOMAS PENNANT'S *TOURS*

R. Paul Evans[1]

In November 1780 the literary wit Horace Walpole critically observed to the Revd William Cole:

> He [Pennant] is not one of our plodders: rather the other extreme: his 'corporal' spirits (for I cannot call them 'animal') do not allow him time to digest anything. He gave a round jump from ornithology to antiquity – and as if they had any relation, thought he understood everything that lay between them.[2]

Walpole was commenting upon Pennant's propensity to travel and write at speed: in his opinion, the succession of topographical works which Pennant produced from the early 1770s onwards were superficial, too hastily flung together, and consequently lacking in scholarly depth. They included his two tours of Scotland, published 1771 and between 1774 and 1776, and his *Tour in Wales* which appeared in two volumes between 1778 and 1783. Walpole's charge that Pennant seldom had time 'to digest anything' will be addressed during the course of this chapter, but what is important to note here is the fact that his contemporaries had detected a distinct change in Pennant's literary output during the 1770s. Prior to this date he had only published works on natural history, his magnum opus being his *British Zoology* which appeared in stages under the auspices of the London-based Cymmrodorion Society between 1761 and 1766. This work gained Pennant critical acclaim and established his reputation as a leading scholar in the field of natural history, especially in ornithology and zoology. It was followed in 1769 with an *Indian Zoology* and in 1771 with the *Synopsis of Quadrupeds*. What contemporaries like Walpole identified, then, was a move away from the production of natural history publications which had so dominated his academic studies during the 1750s and 1760s, towards a series of topographical works which described tours through previously little explored regions of the British Isles such as the Scottish Highlands and the mountains of north Wales. The principal aim of this chapter is to determine at what stage, if at all, it is possible to detect a change of emphasis in Pennant's writing, and if so how and why such a change might have taken place.

The Genesis of Pennant's Interest in Natural History: A Passion for Minerals and Fossils

It seems quite plausible that Pennant's interest in topographical exploration first developed as a consequence of his passion for natural history, the origins of which lie in his fossil-hunting expeditions during the 1750s. Many of these expeditions had a specific scientific purpose in mind, from the search for mineral and fossil specimens to augment his growing collections, to the observation of animals in their native environment and the gathering of field evidence for his ornithological and zoological writings, to viewing a 'cabinet of curiosities' belonging to a correspondent, or sharing mutual interests with other fellow naturalists. Whatever the specific reason, and often it was a combination of factors, natural history field-expeditions dominated Pennant's activities throughout the 1750s, 1760s and early 1770s.

In 1746 or 1747, whilst an undergraduate at the Queen's College, Oxford, he undertook a geological excursion into Cornwall where he met the Revd Dr William Borlase, the man who he later claimed had first instilled in him 'a strong passion for minerals and fossils'.[3] During 1747 he explored the Isle of Wight, searching for fossils in the soft chalk cliffs.[4] In 1750 he visited the Isle of Sheppey, rambles which he later commented were undertaken 'in order to collect the various extraneous fossils with which the cliffs on the north side abound'.[5] In 1752 he travelled to the spa town of Buxton,[6] stopping en route to explore the brine pits and salt mines at Northwich in Cheshire, during which he

> descended thro' a dome, and found the roof supported by pillars, about two yards thick and several in height; the whole was illuminated with numbers of candles, and made a most magnificent and glittering appearance.[7]

During the summer of 1754 he made a 15-week tour of Ireland, the primary reason for the expedition being to collect geological specimens.[8] While waiting to cross to Dublin from Holyhead he had a chance meeting with William Morris, a customs official at the port who was the youngest of the three Morris brothers of Anglesey, two of whom, Lewis and Richard, had been responsible for founding the Cymmrodorion Society in London in 1750. The meeting proved advantageous to Pennant on several fronts. William was a keen naturalist,[9] while Richard would later use his position as a clerk in the Naval Office in London to establish for Pennant a lucrative correspondence involving the exchange of specimens with naturalists across the Mediterranean.[10]

During his Irish tour Pennant kept a daily journal and his brief comments allude to places and sites visited as well as people he met. It was not the informed journal of an author who intended publication, but rather a practical working document which concentrated upon geological interests such as visits to mines and quarries and the provenance of particular specimens. The outline itinerary does, however, occasionally provide specific details. In Dublin, for instance, he spent several days in the company of the Revd Richard Barton of Lurgan with whom he had previously exchanged fossils.[11] Travelling anti-clockwise around the country he continued north, and after Ballycastle visited the Giant's Causeway which, he commented, 'was not so great as I imagined'.[12] (He did still feel it necessary, however, to have a basalt column sent back to his home at Downing.)[13] At Muckross in Co. Kerry he visited the copper mines, and his mineral collection of over 800 specimens (now in the Natural History Museum) contains these and others collected during the Irish expedition.[14] Shortly after Pennant's return to Downing in early October William Morris wrote very excitedly to his brother Richard in London to tell him that he had just received from his Flintshire correspondent:

> flychaid o bethau gwerthfawr anial [a box of extremely valuable things], viz., fossil shells, plants, animals, mines minerals etc. Wala, ni bu erioed wr mwynach ar wyneb y ddaearen hon rwyn llwyr gredu [Well, there never was a nicer man on the face of the earth I fully believe]. There are twenty-six parcels of these curiosities.[15]

When evaluated together, Pennant's journal and his varied correspondence point to a strong geological focus during his expedition across Ireland, and this passion for collecting minerals and fossils reached a peak during the mid-1750s. During June 1755 he explored the Forest of Dean and the banks of the river Severn in search of specimens,[16] and in August of that same year he traversed Snowdonia in the company of William Morris.[17]

Such rambles continued almost unabated until the late 1750s but as the decade drew to a close two factors contributed towards a decline in the extent and scope of his geologizing. In 1759 he married Elizabeth Falconer and the domestic and social responsibilities which this now entailed limited his wanderings. But Pennant's own interests were also now shifting from geology towards ornithology and, to a lesser extent, zoology. By 1762 Pennant could inform the mineralogist Emanuel Mendes Da Costa that he felt he had now 'exhausted' the fossil and mineral kingdom and was now turning his attention to a new branch of natural history; he had set himself the task of compiling a 'British Zoology', the first part of which was to be a catalogue of native birds.[18] However, the writing of such a work required more extensive research and active fieldwork than he had previously undertaken.

The Birth of a Naturalist: A Shift of Interest from Geology to Ornithology and Zoology

Two bereavements, his father in 1763 and his wife in 1764, caused Pennant to slide into a period of melancholy and depression, and it was partly in an attempt to lift his spirits that he planned a Grand Tour of the near Continent. It was a Grand Tour with a difference: not a typical European tour concentrating upon the Classics and the arts, but the itinerary of a working naturalist keen to gather subscribers for his partly written *British Zoology* and to further his research by networking with his peers, viewing their collections and agreeing to the exchange of specimens. Lasting six months, Pennant's continental journey took him through France, Switzerland, Germany and the Netherlands.[19] Much of his time was spent in the company of fellow naturalists: in Paris, for instance, he discussed research interests and viewed the collections of Mathurin-Jacques Brisson, Henri-Louis Duhamel and Georges Louis Leclerc, the Comte de Buffon.[20] A large part of the journey through Switzerland was made in the company of Rodolph Valltravers from whom Pennant received 'a present of several valuable fossils'.[21] At Berne he struck up a profitable friendship with the ornithologist Daniel Sprungli and the anatomist and botanist Albrecht von Haller.[22] Through such associations Pennant was able to establish a web of correspondents with whom he kept in contact over the following decades. Two friendships established in Germany proved to be particularly profitable. At Leiden he met the Dutch naturalist Laurens Theodorus Gronovius, noting in his journal that he found his cabinet to be 'rich in fish and serpents'.[23] Even more significant was the friendship established with a young German naturalist, Peter Simon Pallas, during a visit to The Hague towards the end of the tour. The chance meeting of these two like-minded individuals resulted in Pennant's second publication on natural history, his *Synopsis of Quadrupeds* (1771).[24]

Friendship with Joseph Banks and Peter Simon Pallas

In 1767 Pallas took up a position with the Academy of Sciences in St Petersburg and between 1768 and 1774 he led scientific expeditions to explore the vast expanse of Siberia. Pennant was anxious to be kept informed of

Pallas's travels, requesting information relating to the discovery and recording of fauna of these little-traversed regions.[25] It was during this same period that another of Pennant's new acquaintances, Joseph Banks, began his numerous journeys of exploration and discovery.

Pennant had first become acquainted with Banks in March 1766,[26] possibly through the influence of a mutual friend, John Lloyd, 'the philosopher' of the Hafodunnos and Wigfair estates near St Asaph. Only a month after their initial meeting Banks embarked upon a 10-month expedition to Newfoundland and Labrador. Pennant was deeply fascinated by these explorations and prior to Banks's departure for the northern seas he supplied him with a notebook full of natural history questions which he hoped his new friend would answer.[27] Hearing of Banks's safe return early the following year Pennant wrote in haste that he was 'impatient to survey the treasures you bring home; & rejoice that ornithology is to receive such improvements from yr labors'.[28] In November 1767 Banks visited Downing and stayed with Pennant for 12 days during which time the two naturalists explored the neighbouring countryside and viewed the latter's natural history collections.[29] Pennant returned the visit the year after, travelling to Lincolnshire to spend time at Banks's home at Revesby Abbey.[30]

There followed a three-year gap in their correspondence owing to the fact that in August 1768 Banks, with the naturalist Dr Daniel Solander, set sail with Captain Cook on the *Endeavour* on a voyage of exploration to the South Pacific which would make a significant contribution to the study of natural history. It was not until July 1771 that Pennant heard that Banks had survived the journey and made his way with great haste to London to view the flora and fauna specimens collected in the Southern Hemisphere by his naturalist friends.[31] It was more than likely that these momentous expeditions by his colleagues Pallas, Banks and Solander inspired Pennant to undertake his own journey of discovery, if on a less dramatic scale. In June 1769 he decided to venture north into Scotland.

The Urge to Explore and Discover the Fauna of Scotland

For much of the eighteenth century Scotland was viewed by its southern neighbours as a wild, inhospitable region. The 1715 and 1745 Jacobite Rebellions had resulted in a policy of severe repression and this together with its geographical isolation provided little inducement for English travellers to venture north; as Dr Johnson remarked, to Englishmen the area north of the Tweed was as 'equally unknown with that of Borneo or Sumatra'.[32] One of the few who did explore was Daniel Defoe, who toured Britain between 1724 and 1726. His experiences and hardships forced him to conclude that the Highlands was 'a frightful country, full of hideous desert mountains and unpassable except to the Highlanders'.[33] Anti-Scottish prejudices were strong and, with the exception of Martin Martin in 1703, few travellers before Pennant spoke well of Scotland.[34] Richard Pococke, the bishop of Ossory, made three tours of the Highlands in the years 1747, 1750 and 1760, and though his work was not published in his lifetime he was a correspondent of Pennant's and the latter may well have had a copy of Pococke's manuscripts when writing up his own *Tour*.[35]

Scotland was thus ripe for exploration, and in 1769 Pennant under-took what was considered by many of his friends to be a dangerous and somewhat foolhardy expedition to the Scottish Highlands. This journey would enable Pennant to fill the gaps in his knowledge of the natural history of Scotland, about which very little had been written. In his *British Zoology* Pennant had relied heavily upon the works of Martin and, to a lesser extent, the publications of Sir Robert Sibbald.[36] But he was a firm believer in field observation and, besides being always hopeful of discovering a 'non-descript', wished to study birds and animals in their natural habitat. There can be little doubt that the 1769 tour was made principally for the purposes of fulfilling his study of natural history, a desire clearly identified in the preface:

> I cannot help making this application to myself, who, after publishing three volumes of the Zoology of GREAT BRITAIN, found out that to be able to speak with more precision of the subjects I treated of, it was far more prudent to visit the whole than part of my country: Struck therefore with the reflection of having never seen SCOTLAND, I instantly ordered my baggage to be got ready, and in a reasonable time found myself on the banks of the Tweed.[37]

As the Revd Gilbert White of Selborne commented to Pennant after his safe return home to Downing:

> You must have made, no doubt, many discoveries, and laid up a good fund of materials for a future edition of the British Zoology: and will have no reason to repent that you bestowed so much pains on a part of Great Britain that perhaps was never so well examined before.[38]

That Pennant achieved so much in his 1769 expedition is all the more remarkable given that it was such a hastily arranged venture. In many respects this was in complete contrast to his second tour of Scotland in 1772, which was carefully planned and for a specific reason – to gather a wealth of information for a publishable travelogue.

The Expedition of 1769 – the 'Tour of a Naturalist'

Pennant left Chester on 26 June 1769, just a few days after his 43rd birthday. He proceeded east across country, travelling via Northwich, Macclesfield, Buxton, and Chesterfield to Lincoln and the heart of the Fen country. He spent several days in the Fens engaged in ornithological observation, noting in his journal that 'the birds which inhabit the different Fens are very numerous: I never met with a finer field for the Zoologist to range in'.[39] Proceeding north along the east coast he hired a small boat to transport him to Flamborough Head where he observed nesting sea birds up close:

> The cliffs are of a tremendous height, and amazing grandeur […] the color of these rocks is white, from the dung of the innumerable flocks of migratory birds which quite cover the face of them, filling every little projection, every hole that will give them leave to rest; multitudes were swimming about, others swarmed in the air, and almost stunned us with the variety of their croaks and screams; I observed among them cormorants, shags in small flocks, guillemots, a few black guillemots, very shy and wild, auks, puffins, kittiwakes, and herring gulls.[40]

The following week he took to the sea again to view the Farne Isles:

> Visited those islands in a coble, a safe but seemingly hazardous species of boat, long, narrow, and flat-bottomed, which is capable of going thro' a high sea, dancing like a cork on the summits of the waves. […] Landed at a small island, where we found the female Eider ducks at that time sitting […] The Ducks sit very close, nor will they rise till you almost tread on them […] We robbed a few of their nests of the down, and after carefully separating it from the tang, found that the down of one nest weighed only three quarters of an ounce, but was so elastic as to fill the crown of the largest hat.[41]

It was much to his delight that he could record observing over twenty species of different birds, making this field observation at its best (Fig. 1.1).[42] Continuing north he then crossed the Tweed at Berwick and entered Scotland, arriving in Edinburgh on 18 July, some three weeks after leaving Chester. His journal up to this point contains a heavy emphasis upon natural history observation but after crossing the Tweed we can detect a subtle change in his subject matter. Having now left the prolific bird life of the east coast, his ornithological observations became less frequent; greater attention was paid to the topography, including the introduction of some historical anecdote.

After spending a week in the capital exploring places of interest, Pennant then crossed the Firth of Forth and headed for Perth and the Highlands proper.

Figure 1.1 *Plate I. The Eider Drake and Duck* (drawn by Sydney Parkinson, the artist employed by Banks during his voyage on the *Endeavour*), from *A Tour in Scotland, MDCCLXIX* (Chester, 1771).

Unlike the reports of previous travellers his descriptions of the mountainous landscape he now traversed were very favourable:

> The pass into the Highlands is awefully magnificent; high, craggy, and often naked mountains present themselves to view, approach very near each other, and in many parts are fringed with wood, overhanging and darkening the Tay, that rolls with great rapidity beneath. After some advance in this hollow, a most beautiful knowl, covered with pines, appears full in view; and soon after, the town of Dunkeld, seated under and environed by crags, partly naked, partly wooded, with the summits of a vast height.[43]

Such a description contrasts starkly with the writings of Defoe and others: here we detect a growing appreciation of the landscape, a theme which develops throughout his journey. From Dunkeld he travelled to Taymouth, the residence of the Earl of Breadalbane. He visited Loch Tay and proceeded north via Blair Atholl, Aberdeen, Banff and Inverness, reaching John O'Groats on 21 August, nearly two months into his expedition. Only the harshness of the weather prevented him from taking a boat to the Orkney Isles.[44]

Travelling through this difficult terrain on horseback was often hazardous, as his account of his journey through the pass of Killiecrankie testifies:

The road is most dangerous and the most horrible I ever travelled: a narrow path, so rugged that our horses often were obliged to cross their legs, in order to pick a secure place for their feet; while, at a considerable and precipitous depth beneath, roared a black torrent, rolling through a bed of rock, solid in every part but where the Tilt had worn its antient way.[45]

What particularly caught Pennant's attention during his passage along the far northern shoreline between Duncansby Head and St John's Point was evidence of more extensive fauna, especially the bird life, which included:

multitudes of Gannets, or Soland Geese, on their passage northward: they went in small flocks from five to fifteen in each, and continued passing for hours: it was a stormy day; they kept low and near the shore; but never passed over the land, even when a bay with promontories intervened, but followed (preserving an equal distance from shore) the form of the bay, and then regularly doubled the Capes.[46]

He also recorded observing swans, salmon, seals and Highland cattle.

From John O'Groats Pennant then retraced his steps back to Inverness and proceeded west along the shores of Loch Ness and Loch Lochy. Upon leaving Fort William he passed by the mountains of Glen Coe and travelled through Inveraray, skirting the banks of Loch Lomond. By 8 September he had reached Glasgow, 'the best built of any modern second-rate city I ever saw'.[47] He then travelled via Stirling, Falkirk and Queensbury back to Edinburgh. Departing the capital on the 18th he proceeded south via Carlisle, Penrith, Lancaster, Preston and Warrington, returning to Chester on 22 September.

The tour had lasted just over three months, during the course of which he had travelled over 1622 miles.[48] His party had been small, consisting of only himself and a personal servant – but unlike the 1772 tour, apparently, without an artist. Reference is made within the narrative of the tour to members of the Scottish gentry or their stewards who acted as guides and conducted Pennant through their estate lands, but no reference is made to any other constant travelling companions. It would seem that most nights were spent in local inns or, if they were lucky, the home of a gentleman or clergyman. Pennant was impressed by the consistent hospitality he received throughout his journey, despite the extreme poverty of the people:

the houses of the common people decent, but mostly covered with sods; some were covered both with straw and sod. The inhabitants extremely civil, and never failed offering brandy, or whey, when I stopt to make enquiries at any of their houses.[49]

Reflecting later upon his experience he commented that the Highlanders were:

> hospitable to the highest degree, and full of generosity: are much affected with
> the civility of strangers, and have in themselves a natural politeness and address,
> which often flows from the meanest when least expected. Are excessively inquisi-
> tive after your business, your name, and other particulars of little consequence
> to them: most curious after the politicks of the world, and when they can pro-
> cure an old news-paper, will listen to it with all the avidity of Shakespear's
> blacksmith.[50]

Precisely why Pennant decided to write up his field notes and present them to the
public in the form of a 'tour' is unclear, but various possibilities suggest themselves. When the 1769 *Tour* was published it was dedicated to Pennant's good
friend and near neighbour, Sir Roger Mostyn, with an apology for having been
unable to keep Sir Roger regularly informed of his progress, 'the attention of
the traveller being so much taken up as to leave very little room for the discharge
of epistolary duties'.[51] To make up for this shortfall he hoped that the published
account would prove 'more satisfactory than the hasty accounts I could send you
on my road'.[52] Another, perhaps more compelling reason might be that Pennant
intended the 1769 *Tour* to serve as a supplement to his *British Zoology*, covering
areas previously given scant treatment. This is supported by the various references to fauna throughout the narrative and by the fact that the appendix contains a catalogue of the quadrupeds, birds, reptiles, fish and crustacea native to
Scotland, and that 7 of the 18 plates in the volume are of animals.[53] An additional motivational factor – at least, one presented as such retrospectively – was
Pennant's desire to give an unbiased account of 'a country almost as little known
to its southern brethren as *Kamtschatka*',[54] an account which 'labored earnestly
to conciliate the affections of the two nations, so wickedly and studiously set at
variance by evil-designing people'.[55] This wish to heal political wounds certainly
struck a chord at the time, and initial reviews of the *Tour* recognized the significance of his attempt. In January 1772 the *Monthly Review* concluded:

> It has, for a few years past, been the fashion [...] to ridicule and vilify the Scots
> and Scotland, in the keenest and grossest manner [...] The natives of North
> Britain have been represented [...] and the country itself described as the seat
> of indigence and misery [...] But the more candid, the more gentlemanlike
> writer of the present Tour, gives us a very different idea both of the people and
> of the country, in general, so different indeed, that the perusal of his book is
> sufficient to excite an earnest desire in his readers to make the same excursion;
> and we are verily persuaded that it WILL produce that effect: to the mutual
> advantage, perhaps, of both nations.[56]

Such was the demand sparked by reviews like this that the first print-run was quickly sold out, necessitating the publication of a second edition in 1772, a third in 1774, a fourth in 1776 and a fifth in 1790.

Preparing for a Second Scottish Expedition – 'Desirous of Being at once Directed to the Objects Most Worthy the Observation of a Traveller'

The 1769 *Tour* had been a new departure for Pennant, evoking as it did not only the natural environment he travelled through, but its inhabitants, their settlements and their history. This topographical approach had proved popular with readers, a fact which doubtless helped convince Pennant of the need to undertake a second, more extensive, expedition into Scotland. This tour would delve even deeper, producing a narrative rich in historical anecdote, and discussion of sites of antiquarian interest such as standing stones and burial sites, ruined castles and fortifications, abbeys and churches. Yet, the decision to make a second expedition had been taken prior to the publication and favourable reception of his 1769 *Tour*. Writing to George Paton, a clerk at the Customs House in Edinburgh, in December 1771, Pennant had already confessed to 'a strong inclination to make a second excursion into your country: but in particular the western parts & some of the islands'.[57] Unlike the 1769 *Tour*, Pennant's second expedition was planned in advance, and Paton would prove to be a major contributor to the success of the venture, not only in terms of the preparatory work and the execution of the expedition itself, but also in the later editorial work preparing the final manuscript for publication in 1774 and 1776. In mid-December Pennant requested advice from his Edinburgh correspondent on how best to charter a ship to enable him to visit the Western Isles.[58] With Paton's assistance an open letter was printed in the *Scots Magazine* in April 1772 addressing 'Every Gentleman desirous to promote the publication of an Accurate Account of the Antiquities, Present State and Natural History of Scotland'[59] – the varied range of interests represented here would come to characterize the published tour itself. In essence the letter was a direct (and nicely judged) appeal for assistance, to ensure that maximum benefits would be obtained from the journey:

> The great civility and hospitality I experienced in my journey through part of North Britain in 1769, encourage me to make a visit to the places I have not yet seen. Permit me, to prepare you for my coming, by sending this notice of my intention of being in your neighbourhood the ensuing summer, and of paying respects to you. As my stay can be but short, I am desirous of being at once

directed to the objects most worthy the observation of a traveller, and to be favoured with a collection of such things as I take the liberty of enumerating, and which it is impossible for a transient visitor to get together. As my sole objects are my own improvement, and the true knowledge of your country, hitherto misrepresented, I have no doubt of your complying with my wishes, which are included in the following queries and requests.[60]

With his list of 'queries and requests' Pennant was following in the footsteps of his fellow countryman Edward Lhuyd who, in 1696, had issued over four thousand parochial queries to the clergy in Wales while researching his *Archaeologia Britannica*. Pennant was certainly familiar with Lhuyd's methods of research and may well have been influenced by this in his direct appeal for help. Circulated queries had also been issued by the natural philosopher Robert Boyle and in Scotland by the natural historian and geographer Robert Sibbald. Two lists of these queries were printed as appendices to the 1769 *Tour* and comprised 27 questions relating to the 'antiquities and natural history of their parishes' and 45 questions relating to 'the natural history of the parish' (Fig. 1.2).

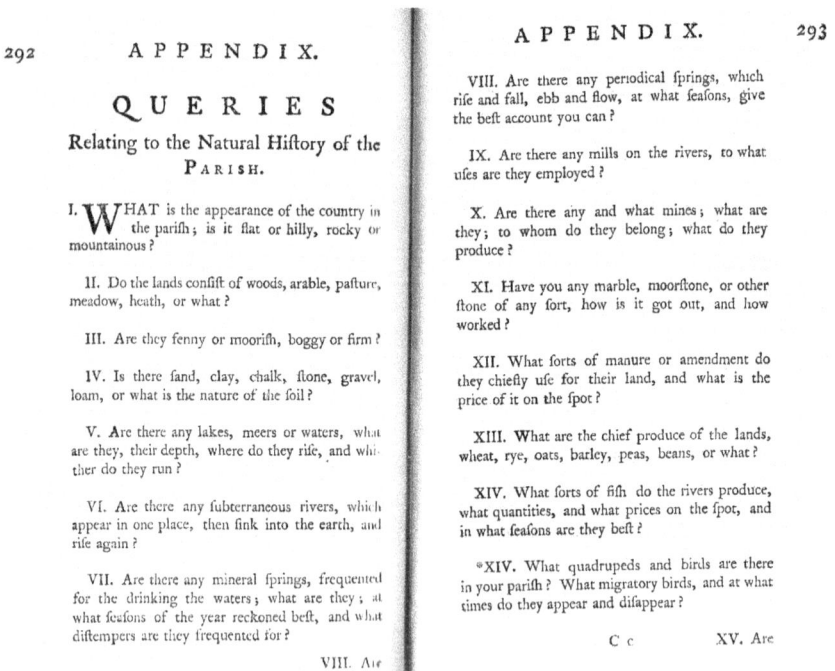

292   A P P E N D I X.

Q U E R I E S

Relating to the Natural History of the
P A R I S H.

I. WHAT is the appearance of the country in the parish; is it flat or hilly, rocky or mountainous?

II. Do the lands confist of woods, arable, pafture, meadow, heath, or what?

III. Are they fenny or moorifh, boggy or firm?

IV. Is there fand, clay, chalk, ftone, gravel, loam, or what is the nature of the foil?

V. Are there any lakes, meers or waters, what are they, their depth, where do they rife, and whither do they run?

VI. Are there any fubterraneous rivers, which appear in one place, then fink into the earth, and rife again?

VII. Are there any mineral fprings, frequented for the drinking the waters; what are they; at what feafons of the year reckoned beft, and what diftempers are they frequented for?

VIII. Are

A P P E N D I X.   293

VIII. Are there any periodical fprings, which rife and fall, ebb and flow, at what feafons, give the beft account you can?

IX. Are there any mills on the rivers, to what ufes are they employed?

X. Are there any and what mines; what are they; to whom do they belong; what do they produce?

XI. Have you any marble, moorftone, or other ftone of any fort, how is it got out, and how worked?

XII. What forts of manure or amendment do they chiefly ufe for their land, and what is the price of it on the fpot?

XIII. What are the chief produce of the lands, wheat, rye, oats, barley, peas, beans, or what?

XIV. What forts of fifh do the rivers produce, what quantities, and what prices on the fpot, and in what feafons are they beft?

*XIV. What quadrupeds and birds are there in your parifh? What migratory birds, and at what times do they appear and difappear?

C c   XV. Are

Figure 1.2 A section from the Appendix of the 1769 *Tour* showing the queries sent out by Pennant to the gentry and clergy of Scotland in 1772, from *A Tour in Scotland, MDCCLXIX* (Chester, 1771).

Key to the preparations was George Paton who, based as he was in the Customs House in Edinburgh, had many useful contacts. In May 1772 Paton notified his Flintshire friend that he had located a Dr Clapperton of Lochmaben near Lockerbie who was 'very knowing in the antiquities & natural History of that part of the country', and to whom he had just posted Pennant's letter and queries. He had also arranged contact with a Mr Maxwell, the Duke of Queensbury's principal steward at Drumlanrig castle who would 'shew all the Beauties of that place'. The same letter provides a list of possible places to visit, including the new abbey at Galloway, the falls on the River Cludden, the 'several Roman stations & remarkable mineral wells' at Moffat, Durisdeer Roman camp and a visit to Iona to 'ascertain the truth of 48 kings being buried there'.[61]

Such informed guidance certainly enabled Pennant to undertake a much more in-depth expedition, as did his decision to take with him a number of learned companions. Chief among these was the Revd John Lightfoot of Uxbridge, whose botanical expertise could bridge Pennant's own shortfall in his knowledge of flora. Also in attendance was his recently acquired painter-servant, Moses Griffith, who would be used to sketch and paint scenes of interest, as well as his French valet, Louis Gold, a fowler and a groom.[62] A last-minute addition was the Revd John Stuart, whose knowledge of Gaelic and 'the customs of the natives' proved invaluable both during and after the tour.[63]

The 1772 Tour of Scotland: 'The Longest of My Journies in Our Island'

Pennant finally left Chester on 18 May and travelled up the west coast of England, through Lancashire into the Lake District. Continuing north through the counties of Dumfries and Lanark he reached Glasgow on 10 June where he was honoured with the freedom of the city. He then sailed down the Clyde to Greenock and there chartered a cutter, the *Lady Frederic Campbell*, captained by Archibald Thomson, to transport him between the isles of the Hebrides. Upon learning of his friend's itinerary the antiquarian scholar Richard Gough commented to the Revd Michael Tyson:

> Mr Pennant is setting out to *approfondir* every corner of Scotland, both land and sea. I verily believe where horses and boats fail, he will take unto himself cork jackets, if not wings.[64]

His voyage lasted a month during which time he visited the isles of Bute, Arran, Gigha, Islay, Jura, Oransay, Colonsay, Iona, Canna, Rum, Skye and

Mull. He noted the flora and fauna of the isles, together with the topography and the history and culture of the islanders, spending some eight days exploring the Isle of Skye. While on the Isle of Arran his attention was directed to the shores of Lochranza to inspect the corpse of a basking shark ('a monster... Twenty-seven feet four inches long')[65] which had been harpooned the previous day (see Fig. 10.2).

Aside from the natural history and the natural beauty of the landscape over which he travelled, Pennant was profoundly struck by the extreme poverty experienced by the inhabitants of the Hebrides, a state they appeared to do little to shake off. On Islay he visited the home of a weaver and was distressed by what he saw (see Fig. 1.3):

> A set of people worn down with poverty: their habitations scenes of misery, made of loose stones: without chimnies, without doors, excepting the faggot opposed to the wind at one or other of the appertures, permitting the smoke to escape through the other, in order to prevent the pains of suffocation. The furniture perfectly corresponds: a pothook hangs from the middle of the roof, with a pot pendent over a grateless fire, filled with fare that may rather be called a permission to exist, than a support of vigorous life: the inmates, as may be expected, lean, withered, dusky and smoke-dried. But my picture is not of this island only.[66]

Inside of a WEAVERS COTTAGE in ILAY   Grignion Sc.

Figure 1.3 *Plate XVI. Inside a weavers cottage on the Isle of Islay,* from *A Tour in Scotland, and Voyage to the Hebrides, MDCCLXXII* (Chester, 1774).

On Rum, he encountered a similar scene of abject poverty, although here his despair was partly alleviated by the warm reception of the family:

> Yet, beneath the roof I entered, I found an address and politeness from the owner and his wife that were astonishing: such pretty apologies! for the badness of the treat, the curds and milk that were offered; which were tendered to us with as much readiness and good will, as by any of Homer's dames, celebrated by him in his Odyssey for their hospitality. I doubt much whether their cottages or their fare was much better; but it must be confessed that they might be a little more cleanly than our good hostess.[67]

To his frustration, Pennant felt that the islanders were so locked into a system of poverty that they lacked the will or motivation to break out of it. The sea off Canna was abundant with fish stocks yet he found that 'the poverty of the inhabitants will not enable them to attempt a fishery'.[68] A similar situation faced the inhabitants of Colonsay:

> Their poverty prevents them from using the very means providence has given them of raising a comfortable subsistence. They have a good soil, plenty of limestone, and sufficient quantity of peat. A sea abounding with fish, but their distressed state disables them from cultivating the one, and taking the other.[69]

What is striking about these observations is the degree to which they differ from the commentary of the 1769 *Tour*. Here Pennant has matured as a social commentator and displays an empathy for the people he meets and a growing consciousness of the landscape over which he travels. Once his voyage was over and he returned to the mainland his description of his journey via Lochs Aw, Lochy, Dochart, Tay and Earn to Perth, particularly attracted the attention of a reviewer in the *Monthly Review* who observed his critical appreciation of the scenery of the Lochs:

> Mr Pennant observes, that it is an idle observation of some travellers with respect to these lakes, that seeing one is the same as seeing all these superb waters; but he shews in a pretty review of all those he has successfully described, that each had its proper and distinct character, and that their appearance are all happily and strikingly varied, to the eye of a nice and judicious observer.[70]

From Perth Pennant continued his journey south via St Andrews, Stirling and Falkirk until he reached Edinburgh on 18 September, where he spent several days exploring the city with his valued correspondent George Paton. Leaving Edinburgh on 26 September he crossed the Tweed at Kelso and then journeyed south via Newcastle, Durham, Halifax, Manchester and Warrington, arriving back in Chester on 12 October. His expedition had lasted five months, during which he had covered an estimated 2,285 miles and undertaken a voyage of over 848 miles.[71]

Preparation and Reception of the 1772 *Tour*

Upon his return to Downing Pennant set about editing his journal ready for publication, a process which took almost four years to complete. While his field notes provided the main skeleton, he relied extensively on additional material from a network of correspondents who had either responded to his earlier 'Queries' or from individuals he had met during the tour itself and with whom he now opened a correspondence. Paton proved invaluable, farming out queries to suitably qualified persons and supplying Pennant with books, journals, maps and drawings. Among those who were thanked in the preface was Joseph Banks, who 'communicated to me his description of STAFFA; and permitted my artist to copy as many of the beautiful drawings in his collection, as would be of use in the present work'.[72] Such was Pennant's gratitude that the 1772 *Tour* was dedicated to Banks:

> I Think myself so much indebted to you, for making me the vehicle for conveying to the public the rich discovery of your last voyage [to Iceland], that I cannot dispense with this address the usual tribute on such occasions. You took from me all temptation of envying your superior good fortune, by the liberal declaration you made that the HEBRIDES were my ground, and yourself, as you pleasantly expressed it, but an interloper.[73]

Among others singled out was his former travelling companion the Revd John Stuart who, in February 1773, promised to supply additional information on the smaller isles of the Hebrides. A detailed account of the Iron Age forts on the Caterthun hills was supplied by George Skene of Careston in Angus, in whose company Pennant had spent a 'day and evening in a most agreeable manner'.[74] In return, Skene remarked that it gave him 'real pleasure to be any ways aiding to you in your researches thro this country, you do us real honour, the country is really obliged to you'.[75] While over 40 individuals were thanked in the preface, what is clear from surviving correspondence is that this was just a fraction of the army of informants who supplied Pennant with specific information relating to their locality.

Such a meticulous approach meant that the first volume of the 1772 *Tour* was not published until 1774 and the second volume, not until 1776. A prompt review in the *London Magazine* acknowledged the pioneering importance of this work which, together with the 1769 *Tour*, served to encourage English travellers to venture north across the Tweed:

Our northern brethren are greatly obliged to him for communicating to the world, the knowledge of their country in its present state, as well as of several ancient customs and manners, and various antiquities, scarcely known before. Our author certainly surpasses all preceding writers on these subjects, and with indefatigable industry hath endeavoured to procure from all parts of Scotland any intelligence that could be of use to the work. The plates also of places most eminent in history, or distinguished by beauty, and of lately discovered antiquities are well executed.[76]

One who quickly followed Pennant's initial tour was the author Dr Samuel Johnson, who, with his companion James Boswell, toured the Highlands during the summer of 1773. Following the publication of Johnson's *A Journey to the Western Islands of Scotland* in 1775 the *Monthly Review* was quick to conclude that Pennant was the more careful observer:

Mr Pennant has led the way; Dr Johnson has followed […] Dr Johnson's book may be regarded as a valuable supplement to Mr Pennant's two accounts of his northern expeditions.[77]

Indeed, Pennant's *Tour* was the occasion for a row between Bishop Percy and Dr Johnson, sometimes referred to as the '*Pennantian Controversy*'. Commenting on Pennant's description of Alnwick castle in Northumberland, Percy concluded that: 'Pennant does not describe well; a carrier who goes along the side of Lochlomand would describe it better.' Johnson disagreed and defended Pennant, claiming him to be: 'The best traveller I ever read, he observes more things than anyone else does.'[78] In a similar incident with Colonel Macleod, who had accused Pennant of superficiality, Johnson retorted: 'Pennant has greater variety of inquiry than almost any man, and has told us more than perhaps one in ten thousand could have done in the time he took.'[79]

Perhaps the most significant aspect of these discussions is the fact that these disputes invariably concerned topographical and antiquarian observations, rather than natural history, thereby illustrating the subtle change in focus in Pennant's writing between his 1769 and 1772 tours.

Addressing the Deficiencies in the *Tour*: 'No Part of North Britain, or Its Islands, Should Be Left Unexplored'

Being conscious of various deficiencies in his own narrative, Pennant set about compiling a map of Scotland and northern England. It was a complex operation which took considerably longer than he expected, with work on it dragging on from 1774 to 1777, partly due to printing delays.[80] He also patronized the work of several individuals and in 1777 paid for the publication of John Lightfoot's *Flora Scotica* which was written using information gathered on their 1772 tour, and to which Pennant contributed a sketch of 'Caledonian Zoology'.[81] During May and June 1776 he paid for the Revd Charles Cordiner of Banff to tour northern Scotland, covering the counties of Aberdeen, Moray, Inverness, Ross and Sutherland. The fruit of Cordiner's travels, a work which bridged natural history and topographical enquiry, appeared in 1780 as *Antiquities and Scenery of the North of Scotland, in a series of Letters to Thomas Pennant, Esqr.*[82] Exploration of the remote islands of Orkney and Shetland was undertaken, again at Pennant's expense, by a local minister suggested by Banks. The Revd George Low of Birsa on Orkney made two excursions during the summer of 1774 and 1778, but for various reasons, and not without the want of trying on Pennant's part, Low's journals were not published within the lifetime of either man. However, Pennant did use some of his material in the introduction to his *Arctic Zoology*.[83]

Not content with the patronage of others Pennant continued with his own exploration of the British Isles, keeping full journals but deciding to publish few of his journeys. In 1773 he toured northern England and in 1774, in company with the draughtsman and topographer Francis Grose, he explored the Isle of Man.[84] In 1776 he visited Warwickshire and the following year Kent, and in 1787, in the company of his son David who had just returned from his first tour of the Continent, he toured the South Coast.[85] During the 1770s he had thoroughly explored his native north Wales and these travels, along with *A Journey from Chester to London* (1782) and *Some Account of London* (1790) were some of the few topographical works he published after his Scottish tours – although several of the others were written up and prepared for publication, Pennant felt they contained too many inaccuracies and hence did not put them into print. Several were published posthumously by his son, including *A Journey from London to the Isle of Wight* and *A Tour from Downing to Alston Moor*, both appearing in 1801, and *A Tour from Alston Moor to Harrogate and Brimham Crags* in 1804. By this time (partly thanks to Pennant himself) topographical literature had become very popular and Pennant's Scottish and Welsh *Tours* had secure reputations as reliable, comprehensive and much-quoted travel guides.

Conclusion: 'I Beg to Be Considered Not as a Topographer but as a Curious Traveller'

Through a comparison of the two Scottish *Tours* of 1769 and 1772 it has been possible to discern distinct changes in Pennant's literary output, supporting Walpole's critical observation that a metamorphosis had taken place. As we have seen, Pennant's early interest in natural history, and the collection of geological specimens led to wider field expeditions to satisfy a growing passion for ornithology and zoology; while writing *British Zoology* (1761–66) and *Synopsis of Quadrupeds* (1771), Pennant realized how little was known about the furthest northern reaches of the British Isles. It was during the course of his two Scottish tours that Pennant began to appreciate other aspects of the landscape and the communities through which he travelled, and to record points of interest in his journal. Successful explorations by colleagues – notably the expeditions of Banks, Solander and Pallas – gave added stimulus for Pennant to push the boundaries of his own journeys. The popular acclaim that followed the publication of his 1769 tour served to encourage him to undertake a more carefully planned and detailed itinerary through the Highlands in 1772. In the words of Horace Walpole, it is indeed possible to detect a 'jump from ornithology to antiquity' in Pennant's literary output.

This does not mean, however, that it is possible to discern a distinct 'break' away from the scientific towards the topographical field of enquiry. The subjects, as was entirely natural in this period, remained very much intertwined. It was, rather, an evolutionary process and in some ways it mirrored the change in contemporary society in the late eighteenth century, which saw a growing awareness and appreciation of the landscape, its inhabitants, their history, tradition and culture. It was, of course, quite common for contemporaries to be interested in natural history and antiquarian studies: Pennant's correspondent and fellow naturalist the Revd Gilbert White was not only the author of *The Natural History of Selborne* (1789) but also *The Antiquities of Selborne* (1789), two works that originally appeared together in one volume.

Perhaps the last word can be left with Pennant himself who, in May 1773, one year before the publication of his second Scottish *Tour*, commented to the Revd George Ashby, a fellow naturalist and antiquarian scholar:

> I beg to be considered not as a Topographer but as a curious traveller willing to collect all that a traveller may be supposed to do in his voyage: I am the first that attempted *travels at home*, therefore earnestly wish for accuracy.[86]

That last point is crucially important – Pennant's desire to make his work as accurate as possible can clearly be seen in the methods he employed to ensure that his second Scottish tour was as well-informed as he could make it. We are all, as a result, the beneficiaries of his endless curiosity.

Notes

1 This chapter is based on a lecture delivered at the University of Glasgow on 1 February 2013; much of the material derives from my unpublished doctoral thesis. See R. P. Evans, 'The Life and Work of Thomas Pennant (1726–1798)' (unpublished PhD thesis, University of Wales, 1993), especially chapter 4.

2   W. S. Lewis, ed., *Horace Walpole's Correspondence with the Rev. William Cole*, 2 vols (Yale and Oxford: Yale University Press, 1937), II: 245, Horace Walpole to William Cole, 24 November 1780.

3   Thomas Pennant, *The Literary Life of the Late Thomas Pennant, Esq., by Himself* (London, 1793), 1.

4   Idem, *A Journey from London to the Isle of Wight*, 2 vols (London: The Oriental Press, 1801), II: 148.

5   Ibid. I: 77. Eleven pages of this published *Tour* (I: 77–88) are given over to describing the various fossil remains found along the cliffs during the 1747 excursion.

6   WCRO, CR2017/TP408/f. 9, Thomas Pennant to Emanuel Mendes Da Costa, Buxton, 29 May 1752. The letter contains a sketch of the rock pits at Northwich.

7   Pennant later included an account of the descent into the salt mine in his *Tour in Scotland 1769* (Chester: John Monk, 1771), 2–3.

8   Reference to Pennant's impending visit to Ireland is made in a letter from E. M. Da Costa expressing the wish that his fellow mineralogist will 'return home safe with your Hibernian spoils for I do not doubt by your diligence & industry you will acquire many Elegant curiosities & make many valuable observations', see WCRO, CR2017/TP408/ f. 104. The letter is undated but its content suggests a date *c*.1754.

9   While waiting for a ship to Dublin Pennant explored the coastal fringes of northern Anglesey with William Morris as his guide, recording in his journal (WCRO, CR2017/ TP18/f. 1):

> Wednesday 26 in the morning visited some rocks near the Head collected some hexagonal crystallizations which incrusted the perpendicular fissures. On the north side of the same Hill found a stratum of brownish white clay [...] Mr Maurice [Morris] acquainted me with the odd vegetation of a round long alga [...] No remarkable shells or fish.

10 One such foreign correspondent with whom Pennant exchanged mineral specimens was Prince Ignazio Biscari (1719–86) of Catania in Sicily; Pennant's letters to him were translated into Italian by Richard Morris at the Navy Office. See NLW, Downing Deeds, MS. 52, ff. i, iv, v and vii.

11 WCRO, CR2017/TP18/f. 1. Barton is mentioned several times in Pennant's letters to E. M. Da Costa in the context of the exchange of minerals and fossil specimens.

12 WCRO, CR2017/TP18/f. 1. The visit took place on Wednesday, 18 July 1754.

13 In a letter to E. M. Da Costa following his return from Ireland, Pennant says the pillar was waiting for him when he arrived back at Downing. See WCRO CR2071/TP408/ f. 106, T. Pennant to E. M. Da Costa, 2 November 1754.

14 W. Campbell Smith, 'The Mineral Collection of Thomas Pennant', *Mineralogical Magazine* XVI (1913): 338. Smith records that Pennant's catalogue of his mineral collection contains lists of specimens collected at Killarney on 22 August 1754, Cobaltite from Muckross copper mine, Co. Kerry, and cerussite from the silver mine district of Co. Tipperary.

15 J. H. Davies, ed., *The Letters of Lewis, Richard and John Morris of Anglesey (Morrisiaid Mon) 1728–1765*, 2 vols (Aberystwyth: privately printed, 1906–8), I: 318.

16 WCRO, CR2017/TP408/f. 126, T. Pennant to E. M. Da Costa, 13 June 1755.

17 In his letter to E. M. Da Costa dated 8 August 1755, Pennant records that his 'journey proved very unprofitable by reason of the extreme badness of the weather. I was more fortunate in anglesea [*sic*] in acquiring a very fine collection of British shells, many unknown

to me before, and as I never thought to be the productions of our shores.' See WCRO, CR2017/TP408/f. 132.

18 In a letter to E. M. Da Costa dated 2 July 1762, Pennant notes: 'having some years ago completing [*sic*] my collection in that Branch of Natural History [minerals and fossils] I have applied myself to another. I formed the great Design of a British Natural History.' See WCRO, CR2017/TP408/f. 165.

19 Pennant kept a detailed journal of his travels on the Continent between February and August 1765, see NLW 12707E; it was later published by G. R. De Beer as *Tour on the Continent, 1765* (London: Ray Society, 1948).

20 Pennant, *Tour on the Continent*, 9.

21 Ibid. 100.

22 Ibid. 103. William Coxe's *Travels in Switzerland* (London, 1791) included a 'Faunula Helvetica' also derived from Pennant's Continental travels: 'The reader is, however, originally indebted to Thomas Pennant Esq., for this catalogue. That ingenious naturalist having communicated to me a list of the Swiss birds drawn from Mr. Sprungli's much-admired cabinet' (v).

23 Pennant, *Tour on the Continent*, 156–57. He had established a correspondence with Gronovius as early as 1761 and it continued until the latter's death in 1777.

24 Idem, *Literary Life*, 7.

25 For their correspondence covering the years 1766–81, see Carol Urness, ed., *A Naturalist in Russia: Letters from Peter Simon Pallas to Thomas Pennant* (Minneapolis: University of Minnesota Press, 1967).

26 Pennant, *Literary Life*, 9.

27 WCRO, CR2017/TP44. The front cover of this notebook is endorsed in Pennant's hand: 'These queries I drew up for Mr. Banks during his voyage to Newfoundland April 1766'.

28 British Museum (Nat. Hist.), D.T.C., vol. I, f. 3, Thomas Pennant to Joseph Banks, 30 January 1767.

29 NLW 147C, ff. 61–65. Banks arrived at Downing on 21 November, as part of a tour from south to north Wales.

30 Pennant, *Literary Life*, 8–9.

31 Ibid. 13.

32 A. J. Youngson, *Beyond the Highland Line* (London: Collins, 1974), 36.

33 Daniel Defoe, *A Tour Through the Whole Island of Great Britain, 1724–26*, ed. Pat Rogers (Harmondsworth: Penguin, 1979), 672.

34 Martin Martin, *A Description of the Western Islands of Scotland* (London, 1703).

35 Richard Pococke, *Tours in Scotland 1747, 1750, 1760* (Edinburgh: Scottish History Society, 1887). When speaking of the hills around Glencoe, Pennant wrote in his journal: 'the hills are very lofty, many of them taper to a point, and my old friend, the late worthy Bishop Pocock, compared the shape of one to mount Tabor'. See Pennant, *Tour in Scotland 1769*, 181.

36 Robert Sibbald, *Scotia Illustrata* (Edinburgh, 1684).

37 Pennant, *Tour in Scotland 1769*, iii–iv.

38 Gilbert White, *The Natural History of Selborne*, ed. R. Mabey (Harmondsworth: Penguin, 1978), 70. The letter is dated Selborne, 8 December 1769.

39 Pennant, *Tour in Scotland 1769*, 10. In a footnote Pennant directs the reader to consult his *British Zoology* in order to obtain 'a more particular account of animals mentioned in this Tour'.

40 Ibid. 15.
41 Ibid. 35–36.
42 Ibid. 36.
43 Ibid. 74–75.
44 Pennant advised future travellers of this route to set off much sooner: 'I found that the bad weather, which begins earlier in the north, was setting in: I would therefore recommend to any traveller, who means to take this distant tour, to set out from Edinburgh a month sooner than myself.' Ibid. 158n.
45 Ibid. 101–2.
46 Ibid. 155.
47 Ibid. 200.
48 This total has been calculated from the figures given by Pennant in his itinerary, ibid. 299–304.
49 Ibid. 66–67.
50 Ibid. 166.
51 Ibid. iv.
52 Ibid.
53 'A Recapitulation of the ANIMALS mentioned in the Tour, with some additional Remarks in Natural History', ibid. 273–86. See also ibid. Plates I, VIII, XIV, XV, XVI, XVII, XVIII.
54 Pennant *Literary Life*, 11.
55 Ibid. 13.
56 *The Monthly Review* (January 1772): 48–49. The *Critical Review* (January 1772) went further and commented that the author was 'justly entitled to the acknowledgement of having obliged the public with the best itinerary which has hitherto been written on that country' (28).
57 NLS Adv. 29-5-5, vol. I, f. 7, letter dated 3 December 1771.
58 NLS Adv. 29-5-5, vol. I, ff. 7 and 8, letters dated 3 and 13 December 1771.
59 *The Scots Magazine* XXXIV (April 1772): 173–75, letter dated 31 January 1772.
60 Ibid. XXXIV (April 1772): 173–75.
61 NLW 2591E, G. Paton to T. Pennant, 18 May 1772. Paton also writes that he would forward further information as he obtained it and address it to Pennant at Carlisle or Glasgow.
62 NLS Adv. 29-5-5, vol. I, f. 12, letter from T. Pennant to G. Paton, 24 April 1772.
63 The Revd John Stuart (1743–1821) was vicar of Killin in Breadalbane and later of Luss in Dumbartonshire. It was he, acting through the intermediary of Paton, who requested permission to accompany Pennant on his tour. See NLS Adv. 29-5-5, vol. I, f. 12; Thomas Pennant, *A Tour in Scotland, and Voyage to the Hebrides, MDCCLXXII*, 2 vols (Chester, 1774–76), iii, Advertisement.
64 John Nichols, ed., *Literary Anecdotes of the Eighteenth Century* VIII (London: Nichols Son and Bentley, 1814), 589.
65 Pennant, *Tour in Scotland 1772*, 192.
66 Ibid. 261–62.
67 Ibid. 317–18.
68 Ibid. 313.
69 Ibid. 273–74.
70 *The Monthly Review* XLVI (January 1772): 153.

71 The total has been calculated from the figures given by Pennant in his itinerary, *Tour in Scotland 1772*, I: 370–73 and II: 469–72. His expense book records a total cost of £296.00, later recouped by selling the copyright of his Scottish *Tours* to his publisher Benjamin White for £550.00 in October 1775. See WCRO, CR2017/TP571, 'Expences of my different works'.

72 Pennant, *Tour in Scotland 1772*, I: iv.

73 Ibid. i.

74 Pennant, *Tour in Scotland and the Hebrides*, II: 165. The account occupies pages 157–60 of the *Tour* and includes a plate showing a ground plan of the White Caterthun Iron Age fort.

75 NLW 15423C, letter from G. Skene to T. Pennant, 29 May 1773.

76 *The London Magazine* 43 (July 1774): 337.

77 *The Monthly Review* LII (January 1775): 57. Horace Walpole went so far as to describe Johnson's Tour as 'a heap of words to express very little', see W. S. Lewis, ed., *Horace Walpole's Correspondence with the Countess of Ossory*, 48 vols (New Haven and Oxford: Yale University Press, 1937–83), I: 225.

78 See J. W. Croker, ed., *Boswell's Life of Johnson, including their 'Tour to the Hebrides'* (London: John Murray, 1890), 587–88.

79 Ibid. 339.

80 See Evans, 'Life and Work of Thomas Pennant', 493–94.

81 John Lightfoot, *Flora Scotica: or a Systematic arrangement in the Linnaean Method of the native plants of Scotland and the Hebrides*, 2 vols (London, 1777).

82 Charles Cordiner, *Antiquities and Scenery of the North of Scotland, in a series of Letters to Thomas Pennant, Esqr.* (London, 1780).

83 For the protracted circumstances surrounding the attempted publication of this work, see Evans, 'Life and Work of Thomas Pennant', 496–501, and the article by C. Stephen Briggs in this volume. For Low's tour, see George Low, *A Tour through the islands of Orkney and Shetland containing hints relative to their modern and natural history collected in 1774*, ed. Joseph Anderson (Kirkwall, 1879).

84 Pennant, *Literary Life*, 22.

85 Ibid. 24–25, 31.

86 Suffolk County Record Office E2/22/2/f. 25, T. Pennant to G. Ashby, 18 May 1773.

Part I

HISTORY, ANTIQUITIES, LITERATURE

Chapter 2

THOMAS PENNANT: SOME WORKING PRACTICES OF AN ARCHAEOLOGICAL TRAVEL WRITER IN LATE EIGHTEENTH-CENTURY BRITAIN

C. Stephen Briggs[1]

Introduction

As Paul Evans has shown in this volume, Thomas Pennant's initial scholarly successes lay in the fields of zoology and Asian geography.[2] Later, however, he was to earn fame and fortune by developing his own literary genre to promote readable, well-illustrated books on exploratory cultural travel in late eighteenth-century Britain. An important legacy to researchers, Pennant's landscape, site and artefact descriptions are unique records of information now often lost. Alongside many other archaeological records, they are being slowly absorbed into searchable, publicly accessible Heritage Environment Records (HERs) throughout Britain; and most pertinently to this investigation, into Wales's 'Coflein', and Scotland's sister database 'Canmore'.[3] These resources not only gather up-to-date data to facilitate archaeological research; they are also tools vital to informing national planning policies, guiding implementation in matters of preserving and conserving sites and landscapes. Adding new data to them demands scholarly judgement to ensure accuracy and reliability. Pennant's archaeological contributions – including his commissioned graphic images – therefore need careful scrutiny to establish the degree to which they truly represented first-hand familiarity with their subject matter.

One of the main purposes of this essay is to consider some of Pennant's working practices as a step towards establishing how such scrutiny may be most usefully progressed. It begins with a brief review of Pennant's education and early mentoring meant to offer insights into his development as an antiquary.

It goes on to include some preliminary observations about his encounters with notable monuments and artefacts and their discoverers. Whereas the graphic records of historic architecture he commissioned and many other aspects of his scholarship merit extended discussion, the present essay is limited to archaeological topics mainly of interest to prehistory. Finally, Pennant's works are evaluated as legacies to scholarship and as documents recording features of a fugitive and continuously fragmenting historic environment before suggestions are offered for future research directions.

Early Influences

Like many of the squirearchy, in youth Pennant would have had access not only to the select library his family had built up at Downing (which he greatly expanded), but also to the collections at nearby Mostyn Hall.[4] In later life he recalled how the Classics were shelved there with 'numerous [...] books related to the Greek and Roman antiquities'.[5] Such advantages must have helped foster in him a basic taste for literature, natural history and antiquarianism before he went up to Oxford. Pennant left college without graduating and travelled intermittently thereafter, as did many of his station.[6] In 1746 or the year after, he visited William Borlase,[7] who had written both an archaeology and a natural history of Cornwall.[8] Having subscribed to and received a copy of the archaeological volume, the young Welshman pledged that he, too, might have 'the courage to attempt the study [...] of antiquity'.[9] Borlase helped develop Pennant's understanding about the periodization of prehistory, encouraged him to collect fossils,[10] and otherwise infected his apprentice with enthusiasms for geology, ancient stone monuments, mining and metallurgy and country houses, all of which were to become important components of his later works.

Aspiring to the company of polite metropolitan learning, Pennant joined the newly established London-based Cymmrodorion Society as a Corresponding Member in 1751, achieving full membership a decade later.[11] His patronage of that Society eventually proved to be advantageous, as in 1766, without much prior evidence of his abilities, it was persuaded to oversee production of his first scholarly publication, the *British Zoology*, which, as its title page declared, was intended to have been 'sold for the benefit of the British Charity-School on Clerkenwell Green'.[12] In the event, the school did not benefit, and given his original claim to have been making a philanthropic gesture, albeit under anonymous authorship, it seems unfortunate that Pennant's later editions do not mention the Cymmrodorion.[13]

While at Oxford Pennant would have learnt how successful scholars usually belonged to one of two institutional camps – occasionally to both. Established in 1660, the Royal Society was the more senior and powerful.[14] Active by 1717, The Society of Antiquaries of London was less well-established. With a fellowship then including a mix of notable eccentrics and accomplished scholars, the Antiquaries promoted historical research of almost every complexion by mid-century. Social exclusivity and apparent lack of direction were, however, features that attracted unremitting public mockery for what some contemporaries considered only trivial achievement. Borlase, a Fellow of the Royal Society by the 1750s, never himself joined the Antiquaries. And though, like his mentor, Pennant was already immersed in natural history, he was still then an undistinguished investigator with little in print. In 1754, ambitious to attain greater recognition in circles of learning and unlikely yet to be elected to the Royal Society, Pennant turned to the Society of Antiquaries.

He was elected on 21 November 1754, aged 28.[15] The ballot paper ensuring his admission read: 'Thomas Pennant Esqr. of Dorening [sic] in Flintshire are recommended by the following Gentlemen to be Admitted a member of this Society, who from their personal knowledge have certified him to be one well vers'd in Antiquities and History, and likely to be a valuable and usefull Member'.[16] He was proposed by Gustavus Brander, James Bernard, James Parsons, Arthur Pond and Henry Baker. All are entered in the ODNB, and curiously, excepting Pond the renowned portraitist, all were distinguished more by achievement in the natural sciences than in antiquarianism. It is unlikely that Pennant joined the Society unaware of its shortcomings, yet fellowship must have looked attractive and socially promising to the young Welsh squire. With its recently granted Royal Charter of 1751 the value of networking within an exclusive society to which peers of the realm were automatically granted fellowship would not have been lost on him.[17]

Pennant's only known contribution to the Society's activities was his donation of a sketch depicting a well-known Middle Bronze Age torc.[18] Exhibited on 13 May 1756, it was entitled: 'Golden Torques from Harlech in Merionethshire A.D. 1692 now in the Possession of Thomas Mostyn of Mostyn in Flintshire, Bart.' and said to be 'the same as that described in Cambden [sic], vol. 2, p. 786' (Fig. 2.1a).[19] The Society's catalogue describes Pennant's image as depicting 'a gold bar torc with bent terminals, distorted from its true shape, from Harlech; and three arrowheads from Ireland, one barbed, and two flakes'. The Society's Minutes further note how the torc had been 'dug up in a garden near' Harlech castle, and described it as 'a wreathed bar of gold (or rather, perhaps, three or four rods jointly twisted) about four feet long'.[20] It was eventually bought at auction from Mostyn descendants by the National Museum of Wales in 1977.[21] Additional notes claim that the accompanying 'five smaller figures [there are only three] [...] were [...] found most frequently near the circular encampments

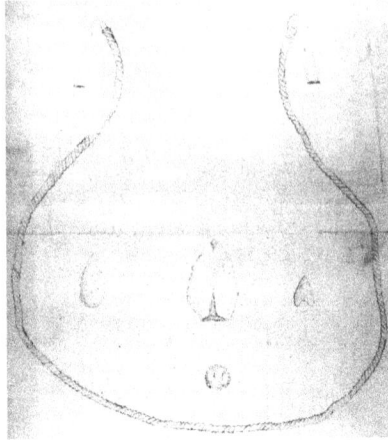

Figure 2.1a *The Harlech Torc*, from The Society of Antiquaries of London Minute Book, 7 May 1761.

[presumably meaning 'raths'] so common in that Kingdom, and called by the Vulgar Irish Elf shots, supposing them to be the [illegible] of Elves and Fairies'.[22] Two of these 'smaller figures' actually depict leaf-shaped hollow-based arrowheads, both probably Neolithic in age. They are placed either side of a diagnostic tool-type known to collectors and scholars since the mid-to-late nineteenth century as 'Bann Flakes', a term locating their distribution to the eponymous river valley in north-east Ireland, where they are thought to have been used for spearing fish during the Mesolithic. As a rare contemporary exhibition of early Irish flint tools, the depiction of this 'Flake' at such an early date is obviously of considerable interest. And as there is no record of an Irish trade in such artefacts at this time, it may be speculated that Pennant himself had collected them when in Ireland only two years before.

Pennant's stay with the Antiquaries was relatively short and he left in April 1761 after only seven years. His resignation claimed 'unwillingness to continue a useless member',[23] a reason said to be owed to the difficulties he had in attending London meetings after his father's death, when he took over the estate. This was a curious argument, however, as the Antiquaries' fees were hardly taxing for a man of his means, and he probably continued visiting London anyway. It may have been relevant that the Society had no regular publishing outlet, since the *Archaeologia* was not established until 1770; a potential reason for his leaving could therefore have been that his publishing aspirations were frustrated, though even when the Society's journal did eventually appear, its mix of courtly deeds, ecclesiastical archival transcripts and medieval tombs may not have been entirely to his taste.[24]

Figure 2.1b *Torques*, by Moses Griffith, from *A Tour in Wales, MDCCLXXIII* (London, 1783).

In 1771, soon after the first *Archaeologia* finally made its appearance, Pennant re-established contact with the Society by presenting another drawing of the Harlech Torc, this time through his friend Daines Barrington.[25] This one was probably from the hand of the squire's recently recruited amanuensis Moses Griffith (Fig. 2.1b) and seems to have carried Pennant's unfulfilled aspiration for the Antiquaries to publish it.[26] Perhaps unsurprisingly, Pennant never sought re-election. Having made an enquiry of the Antiquaries' secretary in 1778, he became incensed at being kept waiting six months for a reply.[27] Probably well out of it, he was elected to the Royal Society only three years later.

Pennant's Tours

Pennant's earliest excursions were not published in his lifetime. Although two small manuscript diaries survive of an Irish visit in 1754, they were never written up because (he claimed) their content was too 'maigre' due to 'the conviviality of the country'.[28] His Continental tour in 1765, moreover, was a diary–travelogue naming every settlement he passed through and giving the most basic descriptions of churches, castles and larger houses; its style offered little foretaste of his later works.[29]

The 1769 Tour to Scotland

Pennant first visited Scotland in 1769. He began in Chester, gave a useful account of the city's antiquities, then crossed to the north-east coast, entering Scotland through Berwick before later exploring the capital. He also visited the Highlands, though on this occasion he never left the mainland. His modest intention had originally been to produce an illustrated guide offering brief topographical histories of ancient features and contemporary landscapes; but though the Romans were strongly in evidence, pre-Roman features were fugitive and artefacts other than medieval weapons few and far between. Antiquities notwithstanding, it is important to note, as Paul Evans shows in this volume, that by regularly incorporating botanical, geological and ornithological observations, all Pennant's topographical texts demonstrated a strong inclination to natural history.

In preparation for this expedition, Pennant had circulated 'Queries addressed to the Gentlemen and Clergy of North-Britain', in order to complement his library research.[30] Since numerous precedents existed for collecting topographical information in this way,[31] he decided initially to employ a reduced version of the 96 questions prepared by James Theobald in 1748, originally compiled for the Society of Antiquaries of London and reprinted by the *Gentleman's Magazine* in April 1755.[32] At first sight it is unclear why Pennant's first version should have reduced the number of Theobald's queries, though as they needed to have particular relevance for Scotland, the attenuation may have been simply to encourage more of its recipients to respond. That he should have then cut them further in advance of the second Scottish Tour[33] might further underline that Pennant's stratagem was indeed to target a wider audience. And to judge from the growing number of his Scottish correspondents during the 1770s and the enthusiastic reception of his first published *Tour*, he was then already making his mark as a successful travel writer.[34] For Scottish readers his crucial insights lay in describing people and scenery without adverse criticism of the Highlanders' primitive lifestyles. One of the first of its kind, Pennant's Scottish *Tour* of 1769 was really an early empathetic ethnography.

The 1772 Tour to Scotland and the Hebrides

The first tour had been something of an ad hoc affair, and upon his return to Downing the squire was reminded in an enthusiastic review that exciting northern terra incognita still remained to be explored – the Hebrides, Orkneys and Shetland.[35] Therefore, only months after completing his first northern journey, Pennant set off on another in the spring of 1772. Now better organized and with many more useful contacts to meet, this time he took a more circuitous, western route through the Lake District.

Castlerigg Stone Circle

As will become clear from this part of Pennant's journey, it is not known how far he stumbled upon, or may have actually prepared to visit his first early monuments before crossing the Scottish border. By the 1770s the antiquarian literature already included some spectacular stone circles in the Lake District. One sited near Keswick was known locally as Castlerigg. It is an unusually extensive monument of around 30 m in diameter which encloses the remnants of an apparently annexed (possibly Neolithic) tomb, not unlike a passage grave. Pennant commissioned 'an excellent drawing by Mr. *John Walker*, of *Keswick*' (Fig. 2.2).[36] The traveller was shown this monument by Dr Brownrigg, a friend and likely collaborator of the Thomas West who authored a *Guide to the Lakes* soon after.[37] Neither Pennant nor his guide were likely to have been aware that this 'Celtic work' had already been discovered by Stukeley as 'Castle-rig', and its internal enclosure described as 'a grave', a half-century before, as the latter's second volume of the *Itinerarium Curiosum* was not printed until 1776.[38] Pennant and Stukeley both narrowly beat Cumbria's first county historians Nicolson and Burn to the press with their descriptions of the site.[39] Ironically, however, by acknowledging 'Doctor *Brownrigg*' as Castlerigg's 'first discoverer', Pennant was the only one to explain how he came to know of its existence.[40]

ANTIQUITIES.

Figure 2.2 Plan of Castlerigg circle, printed from a survey by John Walker. *Antiquities*, from *A Tour in Scotland, and Voyage to the Hebrides, MDCCLXXII* (Chester, 1774).

Concomitant with his later *Tour* of 1772 (1774), Pennant continued publishing additions to his 1769 Scottish *Tour* until a fifth revision of 1790. His addenda included two further groups of Cumbrian prehistoric burial or ritual monuments. The more northerly site, located just outside Penrith, was Arthur's Round Table stone circle and the Mayborough (or Mayburgh) henge monument.[41] Already then well known, this monument complex was soon to be better appreciated from the bird's-eye view published by Stukeley's executors.[42] Pennant himself acquired an accurate plan, here again probably the first printed. Its detail includes a cross section of the henge, an unusual graphic feature for its day.[43] By whom it was surveyed is nowhere stated.

Not far to the south, but north of Shap and close to the routes of the present-day A6 and M6, Pennant encountered 'certain large circles, and ovals formed of small stones' and noted that 'parallel to the road commences a double row of granites of immense sizes, crossed at the end by another row, all placed at some distance from each other'.[44] Though providing no plan on this occasion, he also noted how this monument complex was suffering serious damage. Considerable attrition of unknown extent has certainly occurred since Pennant's day.[45]

Cordiner and Low: Pennant's most northerly informants

When planning his second Scottish excursion in 1772, besides seeing more of the mainland, Pennant decided to visit the Inner Hebrides. Recognizing that he would be unable to cover the most northerly landward counties himself, he 'prevailed on Mr. Cordiner' to collect information there for him. As the Revd Charles Cordiner[46] modelled his approach on Pennant's, the outcome was a published travel journal almost worthy of Pennant's pen and Griffith's pencil.[47] Cordiner's *Antiquities and Scenery of the North of Scotland, in a series of Letters to Thomas Pennant, Esqr.* (1780) involved the same engraver; its research and publication were funded entirely from the Welshman's pocket, and Pennant was proud of it. Though Cordiner hoped his 'antiquities and scenery' would be 'useful appendages' to his paymaster's work, his attachment to Pennant's works was far from slavish. Having examined the monuments for himself, he became seriously concerned at Pennant's rejection of Scotland's brochs and broch-like monuments as places of refuge, defence or domestic occupation: Pennant favoured them as temples for sacred use instead. Cordiner candidly took his would-be mentor to task in print – and at some length – systematically exposed what he considered to be his intellectual naivety.[48] The Welsh tourist seems to have taken it on the chin.

There seemed to be equally poor prospects of Pennant's touring Orkney and Shetland, so in October 1773, acting upon the recommendation of Sir

Joseph Banks, who had been impressed by the excavations of the Revd George Low during his Orcadian visit,[49] he asked Low to report upon the Northern Isles. Low accepted the proposal, Pennant advanced him 'a small sum equal to such a little voyage' and Low travelled north the following May.[50] He was a remarkable self-taught polymath, a 'born naturalist' and had already spent six years in the north preparing a natural history of Orkney. He possessed a fine grasp of the microscope, an unusual understanding of animal and plant ecology and a remarkable eye for archaeology. In executing Pennant's task between 1774 and 1778, Low made exemplary surveys of monuments and drew artefacts and wildlife to a publishable standard. These were passed on in his completed manuscript to Pennant and Gough through their mutual Edinburgh scholar friend George Paton.[51]

Unfortunately, from then on, however, patron and client offered different interpretations of their intentions and histories.[52] It seems worth reflecting at this point upon how this problem was seen in the world beyond Pennant's supportive circle. Writing in 1780, Richard Gough explained that the book had been finished two years before; that its topographical content was quite comprehensive 'with 50 drawings of rude monuments, churches, &c.'. He added 'we are not without hope [it] may shortly be published under the munificent patronage of Mr. Pennant', a turn of phrase all but replicated in his combined high praise and even higher optimism when predicting that 'great improvement in the natural history of Scotland may be expected from the Rev. Mr. Low's Orkney, now preparing for press under [Pennant's] patronage'.[53] Writing in 1786 Thomas Gifford explained how Low had transmitted his 'MS. Account of his voyage' to Pennant, adding that it was 'executed in a very satisfactory manner' and that 'Mr. Pennant mean[t] to complete the voyages of our islands, by publishing, at his own expense, this work of Mr. Low, and should any profits arise, dedicate them to his benefit.'[54]

But as time passed, Gough himself added numerous observations from Low's 'MS. Hist of Orkneys' to a tome he was compiling on sepulchral monuments of all periods – one which became a standard work in its day. In it, he referred to Low's manuscript as if it were already on the library shelves.[55] Pennant similarly employed many of Low's antiquarian observations and half a dozen of his drawings depicting northern antiquities in the first volume of his *Arctic Zoology*.[56] As they hardly related to zoology, it seemed a strange place to put them. Later, both Pennant and Gough vehemently claimed they had tried in vain to get the manuscript published. But after waiting a decade during which his text disappeared from view, it eventually dawned on Low that his promised book was never likely to see the light of day. In 1783 he confessed his despair to Paton: 'As to Monsieur Pennant, I have given up all thought of his patronage, or indeed any body else; and care very little for printing at all.'[57]

By 1788 he could not disguise his anger and depression at the lack of progress and his treatment by Pennant and others:

> but stay [...] what is to be published? Is it not all published already! One has taken a leg, another an arm, some a toe, some a finger, and MR PENNANT THE VERY HEART'S BLOOD OUT OF IT.[58]

By then, Pennant was attempting to blame Gough for hanging on to the manuscript, and in August 1788 suggested to Paton that Low himself was at fault for having mislaid it.[59] It is possible that a more forensic interrogation of all the contemporary correspondence now available could better illuminate this unfortunate episode.

Besides his unique descriptions of unknown ancient monuments, Low had mentioned several prehistoric artefacts. These included 'six pieces of cast brass of a very singular figure [...] seemingly designed for fetters [...] wrapt in a raw hide'.[60] Until recently, their whereabouts was a complete mystery. Then, while researching Scottish Iron Age metalwork at the University Museum of Archaeology and Ethnology in Cambridge, Dr Fraser Hunter of the National Museum of Scotland noticed a certain massive Iron Age armlet. It had been acquired with prehistoric and ethnographic material from the Countess of Denbigh in 1912 when the remnant part of a Pennant–Mostyn–Denbigh collection was finally dispersed. Several of its artefacts were provenanced to Shetland, so with its undoubted Pennant connection and Low's reference to a discovery of 'cast brass[es] of [...] singular figure', Hunter re-established the armlet's Shetland findspot by exemplary detective work.[61] This discovery raises interesting questions about the nature of Low's agreement with Pennant for his Northern Isles commission. The other 'Pennant' items at Cambridge still await study.

The Welsh Tours

Pennant undertook 'two journeys' in Wales between 1773 and 1776 which were published 1778–83.[62] Some of his routes were probably notional and may have incorporated information he had collected on earlier journeys. None of them described Wales south of Montgomery.[63] Nonetheless, together with Pennant's *History of Whiteford and Holywell*, they may be considered to be among the most useful descriptions about Wales's historic environment to have survived from the later eighteenth century.[64]

Pennant was more familiar with the antiquarian topographers of Wales than he had been with those of Scotland. Aside from possessing a fine library of his own,[65] he was also familiar with important libraries holding accessible legacies of Welsh books and manuscripts. The names of Sir John Wynn of Gwydir and Edward Lhuyd were obviously known to him, particularly through the latter's contributions to Bishop Gibson's *Britannia* of 1695 and 1723. As regards other resources, in 1778 Pennant shrewdly borrowed 23 volumes of Lhuyd's manuscripts from Sir John Sebright[66] which the Revd John Lloyd helped translate for him, and in 1776 he hurriedly copied from Thomas Dineley's 1688 manuscript *Tour of Wales* while passing through Badminton.[67]

In Wales Pennant could also make more permanent bonds of friendship with potential local informants living closer to hand, like the Revd John Lloyd of Caerwys and the Revd Richard Farrington of Llangybi.[68] Pennant's lifetime of networking and information-gathering does, however, now pose serious difficulties for establishing whence and from whom some of his information actually came. Those difficulties are briefly addressed here in microcosm through three examples drawn from Pennant's *Tour in Wales*.

The route over Penmaenmawr: Braich-y-dinas

Prior to its existence in near-motorway form, the route of today's A55 was first climbed on foot or horseback, then subsequently by horse-drawn vehicles, eventually including coaches. From medieval times at least, the east–west coastal route across north Wales lay high above its present course, across Penmaenmawr mountain. Today that landscape is dominated by an emptying stone quarry which sits uncomfortably on the seaward plateau rim. It overlooks cairns and stone circles which punctuate the older more inland route. The first known description of Penmaenmawr was written by Sir John Wynn of Gwydir around 1620;[69] Edward Lhuyd included it in his additions to Gibson's edition of Camden's *Britannia* of 1723.[70]

In 1769, Thomas (Governor) Pownall,[71] having read Lhuyd's printed description of Penmaenmawr mountain, was tempted by the prospect of seeing curious antiquities and enjoying the magnificent view, and climbed to the top. The main object of his interest later proved to be the (Iron Age) hill fort, Braich-y-dinas, a site then already known to both Banks and Pennant.[72] The earliest-known graphic evidence of the hill fort comes from Lhuyd's field sketch of *c.*1700: the measurements scribbled upon it suggest the inner circle was 8 paces in diameter, and the outer 35 paces.[73] An identical copy of the plan (though without measurements) is pasted into the NLW grangerized copy of Pennant's *Tour*.[74] That Pennant had access to it confirms that the Lhuydiana he borrowed in 1778 must have included both plans and descriptions of some Welsh monuments seen by Lhuyd and his helpers, though how much of Lhuyd's fieldwork archive came to Pennant in Sebright's mini-archive remains an open question.[75]

Pownall found the hilltop as depicted by Lhuyd, its two smaller concentric stone walls being '7 or 8 feet thick, and about 5 feet high' enclosing a space 'of oval form […] about 30 or 40 yards long, and not quite 20 wide' the innermost wall was 'about 20 feet to the next'.[76] Though Lhuyd's measurements were only approximate, Pownall's are not incompatible with them. These concentric walls appear to have enclosed a long barrow-like twin-peaked cairn.[77] Recognizing that the site's outer stone-built ramparts surrounded hundreds of circular foundations, Pownall dismissed the monument's potential for defence and attributed its foundations 'not to hut footings, but to the *holy compartments* of an extensive Druid temple'.[78]

Pennant obliged Pownall by commissioning a site survey by John Calveley, and 'Captain Grose' was persuaded to engrave an accompanying view. The Antiquaries duly published it.[79] But while bestowing high praise on Pennant for promoting accurate mensuration, Pownall retained his druidical interpretations to the last. Later, in his *Tour* Pennant quietly omitted both Pownall's opinions and their author's name when referencing Sir John Wynn's account and explaining where readers could find a print of the plan he himself had commissioned.[80] After it had been excavated early in the twentieth century, the hill fort was eventually lost to quarrying by 1956.[81]

Richard Farrington was first rector of Llangybi and Llanarmon, later becoming vicar of Llanwnda and Llanfaglan. His serious interest in druidical tradition anticipated Iolo Morganwg by some 30 years.[82] Farrington was also a practising archaeologist who, encouraged by Pennant, had already surveyed a number of megalithic tombs and stone circles before the squire began pursuing his Welsh tour in earnest. His plans are accompanied by sections and isometric views of monuments mostly sited close to the post-medieval west–east routeway of inland Penmaenmawr. Based upon a series of ten itineraries to cairns, circles, standing stones and other monuments and dating from 1769, the most complete manuscript, apparently intended for the press, discusses druidical practice and the druids' uses of their monuments.[83] A second manuscript (of 1772) with similar contents, it lacks the engravings intended for its completion. Their archaeological value has been rehearsed variously in recent times,[84] though their full publication has never been attempted.

Pennant knew Farrington and had encouraged his antiquarian interests before embarking on the *Journey to Snowdon*. Indeed, the tourist narrates how he was taken to Dinas Dinlle and other monuments by the divine. Pennant apparently also enjoyed his hospitality when out exploring. That Farrington bequeathed Pennant 'his entire estate' demonstrates obvious mutual respect between the two men.[85] Interestingly, however, in 1770, probably before Pennant began his Welsh tour in earnest, Farrington had already presented the final draft of his *Snaudonia Druidica or the Druid Monuments of Snowdon* (of

1764–69) to Richard Richardson (of Chester), a relative by his wife with whom he shared a mining venture.[86] And although Farrington dedicated the second of his 'druidical monument MSS' to Pennant in 1772, he moved house to Bath and died the following October. Thus, it seems the squire could not access and inform himself from the illustrated master copy; he also either missed acquiring the version which lacked graphics, or received but felt he could not use it. Farrington's input to the *Tour* was therefore limited by his mortality, and left Pennant to print only minimal evidence of his friend's surveying expertise and goodwill.

Merioneth monuments

Wherever possible Pennant stayed at the houses of county families and minor gentry. In Merioneth, he noted that 'few places […] abound more in British antiquities, than the environs of Cors-y-Gedol', where the Vaughans gave him hospitality.[87] Around 1700 Lhuyd had produced detailed plans of five Neolithic tombs in the area, but (excepting stylized sketches of them that appeared in Stukeley's posthumous *Itinerarium* of 1776), his manuscript versions remained unpublished in Pennant's time.[88] As might be inferred from his printed text, Cors-y-gedol was set in a landscape which then retained far more evidence of ancient settlement than it does today. Though Gibson's *Britannia* speaks minimally of its individual features,[89] Pennant picked out a handful of sites in the locality, namely: Bryn [Bron] y Voel and the Carneddeu [Carneddau] Hengwm megalithic tombs, Craig-y-dinas and Castell Dinas Cortin [Gortin] (hill forts) and the Hengwm Bronze Age stone circles.[90] How he came to know of them is nowhere stated. It could have been from local knowledge, though seems more likely that he was here again informed by Lhuyd's field notes. Because Moses Griffith's unpublished illustration of Carnedd Hengwm in the grangerized *Tour* volume makes little sense, it seems less likely to have been drawn by a draftsman who had ever seen whichever of the two megalithic tombs it was meant to represent,[91] and more likely that Pennant asked his artist to interpret Stukeley's near-incomprehensible later copies of Lhuyd's plans.[92]

The Llanarmon excavation

Paul Evans has explained the central role played by the Revd John Lloyd of Caerwys as friend, scholar–collaborator and translator in the researching of Pennant's Welsh tours.[93] Lloyd's ability to record accurately is well demonstrated by his own account of the excavation he undertook of an early Bronze Age burial at Llanarmon-yn-Iâl (Denbighshire) in which he noted:

June 9th 1774. Employed people to open the remainder of a tumulus in a field called Kefrydd, not far from Tommen y Faerdre, in Llanarmon. When Mr. Pennant and I came there, we found they had demolished one urn that morning; that which was found whilst we were present, was of unburnt clay, full of calcined bones, and undoubtedly bones, ashes &c, converted into more earth; its circumference at the base, two feet two inches, at the projecting part two feet two inches and half, the height eight inches and half. It lay with its face downwards upon a flat rude stone. Over it lay a covering of exceeding fine mould, over that a large flat stone, supported at each end by other stones, to prevent its crushing the urn to pieces. I saw in this tumulus several human bones uncalcined. N.B. Great part of this tumulus had before mixed with lime and carried off for the purpose of manuring the field. By the information of the proprietor of the land, several urns and several skeletons went into the compost.[94]

Regrettably, in his printed tour Pennant omits to mention even that the Revd John Lloyd had been present at the excavation.[95] Furthermore, his friend's explanation that 'several' urns came to light rather than Pennant's minimal two sits uneasily with the more detailed MS account. The urn's present whereabouts is unknown, so its original form can only now be conjectured from the masterly Classical-style image (Fig. 2.3) depicted by Moses Griffith.[96]

Figure 2.3 *The Llanarmon Urn,* from *A Tour in Wales, MDCCLXXIII* (London, 1778).

Conclusions: The Archaeological Legacy of Thomas Pennant

Though there is evidence that Pennant could draw in a workmanlike manner, his discovery that skilled graphic talent could be hired proved of far greater advantage to achieving his objectives than his own skill and first-hand observations ever enabled. It is therefore unsurprising that when comparing Pennant with Gilbert White, the naturalist Richard Mabey felt 'he had no great instinct or aptitude for fieldwork'.[97] As has been demonstrated here, rather than undertake detailed observational fieldwork himself, Pennant preferred to collect and organize information acquired from others. Employing a diverse band of auxiliaries experienced in archaeological and archival research or with exemplary graphic skills enabled the production of topographical works on a scale only otherwise attained on a broader canvas in the eighteenth century by William Stukeley in his own right, and by Edmund Gibson and Richard Gough as successive editors of Camden's *Britannia*. In common with theirs, Pennant's works have inestimable value. His position – which as Paul Evans has documented so well, was focal to a wide network of correspondents – arguably made him a one-man National Monuments Record and came to lend his publications the sort of status later afforded to official Inventories of Ancient Monuments or Victoria County Histories.

Having the means to patronize scholars like Cordiner, Lloyd and Low certainly enabled Pennant bring to light novel, important and often obscure knowledge, helping to make it accessible through publication. Unfortunately, however, Pennant could demand impossible terms and conditions of his collaborators and their contributions, the most obvious casualty being the Revd George Low's remarkable unpublished book. But, again as Paul Evans has clearly demonstrated, Low's case was one among many lacking any acknowledgment at all.[98] Pennant's insensitivity was curiously indiscriminate. And as it made him capable of exploiting even his most powerful allies, it is perhaps fortunate that it took Sir Joseph Banks 'years and years [...] to find out that [he] was to say the least unscrupulous'.[99] Because Pennant thanked and named his informants so selectively, the contents of some of his publications still raise important questions of acknowledgment, accuracy and authority.

Although not always explained in his biographies or obituaries,[100] it appears that even in his day, some of Pennant's activities attracted controversy. He was taken to task in the *Caledonian Mercury* of 10 May 1773, for suggesting in his Scottish *Tour* that 'the immortal Alexander Robertson [...] died as he lived [...] a most abandoned *sot*'; more serious was an article printed in the *Newcastle Courant* on 5 August 1775 alleging that although described in his *Tour*, Pennant had never even seen Alnwick castle. In similar vein, one proud Scotsman felt obliged to defend the history of the MacGregor clan against Pennant's alleged 'libel injurious to an innocent set of people'.[101]

A less adverse outcome of Pennant's first Scottish tour was that by 1776 some northern land agents were advertising those beautiful views he had evangelized, as selling points for rural property.[102] Most literary reviews of his work were complimentary: Samuel Johnson's encomium is well known,[103] while a more humble journalist felt that in his 1769 *Tour* this 'ingenious traveler [had] described numerous scenes [...] in a faithful and entertaining manner'.[104] And in his *Sketch of a Tour in Derbyshire*, Bray says that Mr Pennant 'has an eye to observe, a pen to describe, and a pencil to delineate every thing worthy of observation, in every place to which he comes'.[105] For many of his readers then, Pennant's books were generally felt to be a good read.

There is no doubt that Pennant was a pre-eminent and systematic collector of information and master of scholarly synthesis. Given that many followers who read his books were attracted to Wales and attempted to emulate his literary genre, he certainly deserves the accolade 'Father of Cambrian Tourists'; the *Tours in Wales* continued to be printed or re-edited throughout the nineteenth century. What is perhaps less well known is that in 1808 Richard Fenton the Pembrokeshire antiquary announced his intention of soon 'put[ting] to the press' his county history 'as part of a general Description of South Wales, to form a companion to Mr Pennant's Account of North Wales'.[106] It was also then revealed that Fenton intended 'a new and enlarged edition, in three quarto volumes, of Mr Pennant's Wales'. Fenton was to make appropriate novel explorations with Sir Richard Colt Hoare as illustrator, though drawings would also be forthcoming from Pennant's own collection.[107] A contract between Fenton and 'Messrs. Longman & Co.' was drawn up accordingly.[108] Unfortunately, however, David Pennant probably had a similar project in mind and objected strongly to the proposal, reasoning that he did not wish to see any interference in his father's original text.[109] While editing Fenton's Pembrokeshire *Tour* a century on, his grandson Farrar mistakenly attributed the plan's failure to the older Pennant, rather than to his son David.[110] It seems Fenton's great ambition had been to write a comprehensive history of Wales, county by county, 'but [...] left it rather late in life to tackle such a task'.[111] Colt Hoare's later successes as artist, scholar, tourist and promoter of Welsh county histories probably more than made up for the Pennants' unfulfilled ambitions.[112]

Several tasks now need addressing to help more accurately gauge the value of Pennant's impact upon the history and recording of antiquarian topography and antiquities in eighteenth-century Britain. Whereas the present essay has considered a limited number of the more obvious prehistoric sites and finds mentioned in his texts, his works not only provide plenty of material to expand upon it; they also offer fertile fields for researching Roman, medieval and later topics. Indeed, today those aspects of his texts recording built historic structures could also be re-examined to the considerable profit of historical scholarship.

Future investigative projects might therefore usefully aim to locate all Pennant's unknown or lost early sites and to re-provenance the artefacts he mentioned using relational databases and GIS techniques. Such investigations should be based on the exhaustive listing and analyses of all his (and his collaborators') descriptive texts – including their condition when observed – and should include any related images of the sites and discoveries he recorded. These tasks would ideally involve those charged with collecting Heritage Environment Records and National Monuments Records, thus enabling their ready integration into all the appropriate public databases to greatest advantage. Such an achievement would make a practical and lasting memorial to Pennant's endeavours.

Notes

1 The writer acknowledges the help given unstintingly by Trevor Cowie and Dr Fraser Hunter, NMS; Dr Paul Evans, Ruthin; Ms Caroline Kerkham, Llanddeiniol; the late Gwyn Walters, Aberystwyth, formerly of NLW; Carol Evans and Adrian James at the library of the Society of Antiquaries of London: finally, Dr Scott Lloyd of RCAHMW and the late W. E. (Bill) Griffiths, formerly of that place.

2 See R. Paul Evans's contribution to this volume, 15–37; Charles W. J. Withers, ODNB Online *s.n.* Pennant, Thomas (1726–1798) (accessed 1 February 2015). Withers mentions that the National Library of Wales holds Pennant 'corresp. with Edward Lhuyd'. As Lhuyd died in 1709, this must relate to an indirect connection with Pennant.

3 National Monuments Record Wales, www.coflein.gov.uk; National Monuments Record Scotland, www.rcahms.gov.uk/canmore.html.

4 Eluned Rees and Gwyn Walters, 'The Library of Thomas Pennant', *London Bibliographical Society* (June 1970), 136–49; Eluned Rees, 'An Introductory Survey of Eighteenth-Century Welsh Libraries', *Journal of the Welsh Bibliographical Society* 10 (1971): 197–258 (202).

5 Thomas Pennant, *A Tour in Wales MDCCCLXXIII* (London, 1778), 11.

6 R. Paul Evans, 'Thomas Pennant (1726–98): The Father of Cambrian Tourists', *Welsh History Review* 13 (1987): 395–417.

7 Thomas Pennant, *The Literary Life of the Late Thomas Pennant, Esq., by Himself* (London, 1793), 1.

8 Borlase's publications include *Observations on the Antiquities, Historical and Monumental, of the County of Cornwall* (Oxford, 1754; 2nd ed., London, 1769), and *The Natural History of Cornwall* (Oxford, 1758).

9 P. A. S. Pool, *William Borlase* (Truro: Royal Institution of Cornwall, 1986), 149.

10 For periodization, see P. Rowley-Conwy, *From Genesis to Prehistory and the Adoption of the Three Age System* (Oxford: Oxford University Press, 2007), 113; for fossils, see 'The Pennant Collection', *Science*, 14 March 1913: 404–5.

11 *Gosodedigaethau Anrhydeddus Gymdeithas y Cymmrodorian yn Llundain, Dechreuedig ym Mis Medi, 1751* (Llundain, 1755); *Cofrestr o Gymdeithas y Cymrodorion yn Llundain: Gwyl Ddewi 1762* (Llundain, 1762), 1, where Pennant is described as 'bonheddig' (a gentleman).

12 Its details are worth repeating in full: *The British Zoology: published under the inspection of the Cymmrodorion Society, instituted for the promoting of useful charities, and the knowledge of nature, among the descendants of the ancient Britons; illustrated with one hundred and seven copper plates* (London, 1766).

13 Apparently 'no benefit ever accrued to the School from the sales, though Pennant afterwards gave it a donation of £100', R. T. Jenkins and Helen M. Ramage, *A History of the Honourable Society of Cymmrodorion 1751–1951* (London: Honourable Society of Cymmrodorion, 1951), 82–83.

14 There appears to be no up-to-date history of the Royal Society. For a useful background source, see Richard Sorrenson, 'Towards a History of the Royal Society in the Eighteenth Century', *Notes and Records of the Royal Society of London* 50, no. 1 (January 1996): 29–46. For the Society of Antiquaries of London, see Susan Pearce, 'Antiquarians and the Interpretation of Ancient Objects, 1770–1820', in *Visions of Antiquity: The Society of Antiquaries of London 1707–2007*, ed. eadem (London: The Society of Antiquaries of London, 2007), 147–71, passim.

15 Pearce, 'Antiquarians and the Interpretation of Ancient Objects', 155–56.

16 SAL, Manuscript Minutes VII (18 July 1754): 143.

17 R. B. Pugh, 'Our First Charter', *Antiquaries Journal* 62 (1982): 347–55.

18 SAL, 'Primeval Antiquities' drawing, 37.2.

19 Noted by Lhuyd in Edmund Gibson, ed., *Britannia* (1695), cols 658–59; ibid. 2nd ed. (1723), col. 786. For Pennant's sketch, see Fig. 2.1a, *The Harlech Torc*, The Society of the Antiquaries of London Minute Book, 7 May 1761, f. 335.

20 See 18n.

21 Hubert N. Savory, *Guide Catalogue of the Bronze Age Collections* (Cardiff: National Museum of Wales, 1980), Plate VIIb, no. 301, 126.

22 See 18n. Further information is inserted in the MS here. It includes a reference to an undated 'Method of Fossils' by John Woodward (1665–1728), where his Attempt towards a natural history of the fossils of England [...] in a catalogue of his own collection (London, 1729) was probably intended. That volume included three figures (on page 43): one depicting prehistoric stone axes and another showing six arrowheads which are among the earliest graphics of British later prehistoric stone artefacts.

23 SAL, Manuscript Correspondence, 266 (15 April 1761); SAL, Manuscript Minutes 1761, 335, 7 May 1761.

24 See Pearce, 'Antiquarians and the Interpretation of Ancient Objects', 154.

25 SAL, Manuscript Minutes, 12 December 1771; Pearce, 'Antiquarians and the Interpretation of Ancient Objects', 153, Fig. 50, 156.

26 It was eventually engraved in Thomas Pennant, *A Journey to Snowdon* (London, 1781), Plate II, 133, described 132–34, and reprinted in *Tour in Wales, MDCCLXXIII* (London, 1783), Plate II, opp. 133.

27 Rosemary Sweet, *Antiquaries: The Discovery of the Past in Eighteenth-Century Britain* (London: Hambledon and London, 2004), 64.

28 For the diaries, see WCRO, CR 2017/TP18/ff. 1–2. Pennant refers to his Irish tour in *Literary Life*, 2; see also the discussion by R. Paul Evans in this volume.

29 Gavin de Beer, ed., *Tour on the Continent 1765* (London: Ray Society, 1948).

30 Thomas Pennant, *A Tour in Scotland 1769* (Chester, 1771), 287–98, Appendix III.

31 Adam Fox, 'Printed Questionnaires, Research Networks and the Discovery of the British Isles, 1650–1800', *Historical Journal* 53 (2010): 593–621.

32 For the origins of Theobald's questionnaire, see Joan Evans, *A History of the Society of Antiquaries* (Oxford: Oxford University Press, 1956), 123; for its publication in full, see the *Gentleman's Magazine* 25 (April 1755): 157–58, though contra Fox, 'Printed

Questionnaires', 30, this edition of the *Gentleman's Magazine* is unlikely to have been published by the editor Edward Cave, as he had died the previous January. Pennant quite properly acknowledged his questionnaire's origins in the later, Chester edition of the 1769 *Tour to Scotland*, 287–98.

33 As noted in the *Scots Magazine*, 1 January 1772: 23. At this point Pennant also added a query, making his first total 72: cf. Fox, 'Printed Questionnaires', 31–32. Pennant's final questionnaire appears in the *Scots Magazine*, 1 April 1772: 173–74.

34 Anon., review in *Scots Magazine*, 1 January 1772: 24–27.

35 'Philanthropus', 'A New Tour through Scotland Recommended', *Scots Magazine*, 1 January 1772: 19–21.

36 Plan of Castlerigg circle, printed from a survey by John Walker, *A Tour in Scotland, and Voyage to the Hebrides, MDCCLXXII*, 2nd ed., I: 38, Plate I, no. 1, *Antiquities*. For Castlerigg, see Aubrey Burl, *A Guide to the Stone Circles of Britain, Ireland and Brittany*, revised ed. (New Haven and London: Yale University Press, 2005), 82–83. It has been suggested that Pennant painted a fine watercolour view of Castlerigg in the Bodleian Library, Oxford (Tom Clare in *Prehistoric Monuments of the Lake District* (Stroud: Tempus, 2007, Fig. 6 and 26–27)). The attribution seems unlikely, as Pennant is not believed to have possessed the skill. Two eighteenth-century plans of the site do survive in the Bodleian, however. Archived in Gough Maps 4 (Cumberland), though f. 13b is a crude plan of 1725 by William Stukeley, f. 13a, signed by John Walker, is remarkably similar to the one Pennant printed. I am much indebted to Bernard Nurse, FSA, for kindly drawing them to my attention.

37 Thomas West, *A Guide to the Lakes in Cumberland, Westmorland and Lancashire* (London, 1778), 111–13; ibid. 2nd ed. (Kendal, 1780) 109n. See Burl, *Guide to the Stone Circles*, 42–43. For William Brownrigg, see Herbert T. Pratt, ODNB Online *s.n.* Brownrigg, William (1711–1800) (accessed 31 January 2015). The suggestion of Brownrigg's collaboration with West is made in his biography on 'Wikipedia', http://en.wikipedia.org/wiki/William_Brownrigg (accessed 31 January 2015).

38 William Stukeley, *Itinerarium Curiosum* (London, 1724; 2nd ed. London, 1776); facsimile of the 2nd ed. by Gregg International (Farnborough, 1969), II: 48.

39 J. Nicolson and R. Burn, *The History and Antiquities of the Counties of Westmorland and Cumberland*, 2 vols (London, 1777), II: 80.

40 Pennant, *Tour in Scotland 1772*, 5th ed. 2 vols (London, 1790), 43.

41 Pennant first mentions Mayborough (with plan) in *Tour 1769*, 3rd ed. (Warrington, 1774), Plate XIX, 256, and then ibid. 4th ed. (London, 1776), Plate XXXVI, 276. See also his *Additions to Quarto Edition* (London, 1774), 45. For a useful antiquarian bibliography of the site with discussion of Pennant's plan, see Clare, *Prehistoric Monuments*, 14–15, 56–63; Burl, *Guide to the Stone Circles*, 81–82. For the mid-seventeenth-century plans of Mayburgh circle and henge, see John Fowles and Rodney Legg, eds, *Monumenta Britannica [or a Miscellanie of British Antiquities]: John Aubrey (1626–97)*, 2 vols (Sherborne: Dorset Publishing Co., 1980), 113–14.

42 For the Stukeley view, see *Itinerarium Curiosum* (1776), II: 43 and Plate 84.

43 For discussion of contemporary archaeological surveying, see C. Stephen Briggs, 'From Aubrey to Pitt-Rivers: Establishing a Survey Standard for the British Isles 1660–1860', in *Histories of Archaeological Practices: Reflections on Methods, Strategies and Social Organisation in Past Field Work*, ed. Ola Wolfischel Jensen (Stockholm: National Historical Museum, 2012), passim, 88, Fig. 3.

44 See Thomas Pennant, *Additions to the 1769 Tour*, 3rd ed. (1774), 46; *1769 Tour*, 3rd ed. (Warrington, 1774), 277, where he considers it to have been 'a *Danish* monument'.

45 Aubrey Burl, *From Carnac to Callanish: The Prehistoric Stone Rows and Avenues of Britain, Ireland and Brittany* (New Haven and London: Yale University Press, 1993), 100–101; idem, *Guide to the Stone Circles*, 80; Clare, *Prehistoric Monuments*, 80–83.

46 Jeffrey R. Smitten, ODNB Online *s.n.* Cordiner, Charles (1746?–1794) (accessed 31 January 2015).

47 Cordiner's tour carried 21 illustrations, of which 7 were land- or coastscapes; 5 castles; 3 ecclesiastical; 2 Pictish stones; 2 of a broch; 1 earthwork castle and 1 harbour.

48 Charles Cordiner, *Antiquities and Scenery of the North of Scotland, in a series of Letters to Thomas Pennant, Esqr.* (London, 1780), 104–11.

49 For Low's excavations, see an 'Extract of a Letter from the Reverend Mr. George Low, to Mr. Paton, of Edinburgh, Communicated by Mr. Gough', *Archaeologia* 3 (1773 [1775]): 276–77. For his other collaborations with Banks, see Averil Lysaght, 'Note on a Grave Excavated by Joseph Banks and George Low at Skaill in 1772', *Proceedings of the Society of Antiquaries of Scotland* 104 (1971–72 [1974]): 285–89; idem, 'Joseph Banks at Skara Brae and Stennis, Orkney, 1772', *Notes and Records of the Royal Society of London* 28, no. 2 (April 1974): 221–34.

50 Sweet, *Antiquaries*, 65–66. The history of Pennant's relationship with Low is told through the correspondence between Low, Paton and Pennant (1772–88) by Joseph Anderson, *George Low, A Tour through the Islands of Orkney and Schetland, containing hints relative to their ancient modern and natural history collected in 1774* (Kirkwall: William Peace, 1879), xiv–lxxiv. The book had long been out of print when re-edited by John Hunter (Inverness: Melven Press, 2005). Hunter radically condensed Anderson's valuable introduction, which is now made accessible through a Kessinger facsimile reprint (s.l. [USA], 2014). For Pennant's commissioning of Low, see Anderson, *George Low, A Tour*, xxvii–xxxiii.

51 Pennant, *Literary Life*, xvii.

52 Sweet, *Antiquaries*, 65–66; Anderson, *George Low, A Tour*, passim.

53 Richard Gough, *British Topography: or, an historical account of what has been done for illustrating the topographical antiquities of Great Britain and Ireland*, 2 vols (London, 1780), II: 632, 727.

54 Thomas Gifford, *An Historical Description of the Zetland Islands. By Thomas Gifford, Esqr.* (*Bibliotheca Topographica Britannica* no. 37, V, part 7), xiv.

55 Richard Gough, *Sepulchral monuments in Great Britain applied to illustrate the history of families, manners, habits, and arts*, 2 vols (London, 1786–96). In ibid. I: xi–xiii, Low's manuscript is cited under 'Tumuli in Orkney'; I: x, 'Mr. Lowe, Ms Hist. of Orkney'; I: xii, 'Mr. Lowe' thrice; I: xvii, xxix and lxxix, 'Mr. Lowe'; I: 254, 'Lowe's MS'; II: ccxxxi; 'Mr. Lowe (MS).

56 Thomas Pennant, *Arctic Zoology*, 2 vols (London, 1784–85), I: xxxi–xlix.

57 Anderson, *George Low, A Tour*, lxvi.

58 Sweet, *Antiquaries*, 65–66; Anderson, *George Low, A Tour*, lxix.

59 Anderson, *George Low, A Tour*, lxviii.

60 Ibid. 166–67.

61 Fraser Hunter, 'New Light on Iron Age Massive Armlets', *Proceedings of the Society of Antiquaries of Scotland* 136 (2006): 135–60, 141–48. MS copy of listing of the Pennant Collection, receipt addressed to the Countess of Denbigh, 14 September 1912, Newnham Paddox, Lutterworth. MS signed in the hand of Anatole von Hügel, Curator, University Museum of Archaeology and Ethnology, Cambridge. I am grateful to Drs Fraser Hunter and Brendan O'Connor for this information. The survival of this relict cache of specimens is of particular interest, as on 19 February 1864, fire is said to have destroyed 'most, if not all, of the rare and valuable library of Welsh records

and manuscripts known as the "Mostyn Collection"' then at Pengwern Hall (*Liverpool Mercury*, 22 February 1864; Supplement, *Illustrated London News*, 5 March 1864).

62 Thomas Pennant, *A Tour in Wales MDCCCLXXIII*, I (London, 1778); idem, *A Tour in Wales MDCCLXXVI*, II (London, 1783 [1781]).

63 Evans, 'Thomas Pennant (1726–1798)', 398. 'From Downing to Montgomery and Shrewsbury' is the furthest south in Wales Pennant ventured in print.

64 Thomas Pennant, *The history of the parishes of Whiteford and Holywell* (London, 1796).

65 Rees and Walters, 'Library of Thomas Pennant', passim.

66 Eidem, 'The Dispersal of the Manuscripts of Edward Lhuyd', *Welsh History Review* 7 (1974): 148–78 (162); Evans, 'Thomas Pennant (1726–1798)', 400.

67 For Badminton, see Evans, 'Thomas Pennant (1726–1798)', 411 n. 59, which refers to NLW 2568E. Additionally, NLW 2539C includes three folios in Pennant's hand of notes taken from Dineley's manuscript Tour, therein referred to as Badminton 25. Privately printed in 1864, this was later published as Thomas Dineley, *The Account of the Official Progress of His Grace Henry, the First Duke of Beaufort, Through Wales in 1684*, ed. Robert W. Banks (London: Blades, East & Blades, 1888). Pennant used Dineley's illustrations of Mostyn Hall and the water wheel at Mostyn collieries in his *Whiteford and Holywell* (Figs VI and XVI).

68 R. Paul Evans, 'Reverend John Lloyd of Caerwys (1733–93): Historian, Antiquarian and Genealogist', *Flintshire Historical Society Journal* 31 (1983–84): 109–24; B. G. Charles, 'Some Unpublished Material of the Reverend Richard Farrington', *Journal of the Welsh Bibliographical Society* 4 (1937): 16–32; idem, 'The Reverend Richard Farrington, Vicar of Llangybi,' *Journal of the Welsh Bibliographical Society* 4 (1940): 241–42; NLW 18606C, W. G. Griffiths, 'A Short Memoir of the Revd. Richard Farrington, M.A., 1702–1772'.

69 Sir John Wynn, *An ancient survey of Pen Maen Mawr, North Wales, from the original manuscript of the time of Charles I*, ed. J. O. Halliwell (London, 1859); reprinted by J. Bezant Lowe (Llanfairfechan: W.E. Owen, 1906). The discussion which follows draws on the 2009 pamphlet by T. P. T. Williams which reprints and re-edits Wynn's account with valuable additions, www.worldcat.org/title/ancient-survey-of-pen….

70 Gibson, *Britannia* (1723), cols 804–5.

71 Eliza H. Gould, ODNB Online *s.n.* Pownall, Thomas (1722–1805) (accessed 1 February 2015); Briggs, 'From Aubrey to Pitt-Rivers', 92–93.

72 Governor Pownall, 'Description of the Carn Braich y Dinas, on the Summit of Pen-maen-mawr, in Caernarvonshire', *Archaeologia* 3 (1771 [1775]): 303–9; 303 (306 n. [e] and 309, where it is noted that Pennant intended 'to have an actual survey made of it, as well as of some other places of a like nature, in these parts').

73 BL, Stowe 1023, f. 117/233.

74 See https://www.llgc.org.uk/digitalmirror/jts/JTS00001/index.html?lng=en, for online access to this illustration in the Pennant grangerized *Welsh Tours* volume [for which no NLW MS numbers are offered within the above website], at equivalent of Pennant, *Journey to Snowdon*, 319.

75 It must have been a substantial quantity, as *The Tour in Wales*, 2 vols (London, 1784), contains no fewer than 47 Sebright references in the footnotes, 13 extra references being loosely to Lhuyd by name and possibly meaning the *Archaeologia Britannica* (1707). See Rees and Walters, 'The Dispersal of the Manuscripts of Edward Lhuyd', 162–63.

76 Pownall, 'Description of the Carn Braich y Dinas', 305.

77 Ibid. 304–6, Plate XIV.

78 Ibid. 307–8.

79 Idem, 'Further Observations on Pen-maen-mawr in a Letter to Mr. Gough', *Archaeologia* 3 (1774 [1775]): 350–54.

80 Pennant, *Journey to Snowdon*, 320–21.

81 RCAHMW, *An Inventory of the Ancient Monuments in Caernarvonshire: Volume I: West*, (London: HMSO, 1956), *s.v.* Dwygyfylchi.

82 Charles, 'Some Unpublished Material'.

83 Ibid.; NLW 1118 and 1119.

84 See, e.g., Richard Kelly, 'The Probable Sites of Some Disappeared Chambered Tombs in Caernarvonshire in the Light of Antiquarian Research', *Archaeologia Cambrensis* 123 (1974): 175–79; C. Stephen Briggs, 'Druids' Circles in Wales: Structured Cairns and Stone Circles Examined', *Landscape History* 8 (1986): 5–13.

85 For Farrington's dedication to Pennant, see Charles, 'Some Unpublished Material', 24–25.

86 See NLW 4899B.

87 Pennant, *Journey to Snowdon*, 119.

88 C. Stephen Briggs, 'A Megalithic Conundrum: The Pedigree of Some William Stukeley Illustrations', in *The Founders' Library University of Wales, Lampeter, Bibliographical and Contextual Studies: Essays in Memory of Robin Rider: Trivium*, ed. C. William Marx (Lampeter, 1997), 195–214 (196). The relationship between Lhuyd's manuscripts and Stukeley is discussed most recently in C. Stephen Briggs, 'Meini Gwyr in History and Archaeology', in *Essays Presented to Frances Lynch*, ed. William J. Britnell and Robert J. Silvester (Cambrian Archaeological Society Monograph, 2012), 122–44.

89 Gibson, *Britannia* (1695 and 1723), passim.

90 Pennant, *Journey to Snowdon*, 119–21.

91 https://www.llgc.org.uk/digitalmirror/jts/JTS00001/index.html?lng=en, NLW grangerized volume of the Tour in Wales.

92 Stukeley, *Itinerarium Curiosum* 1776, II: Plate 94.

93 Evans, 'Reverend John Lloyd of Caerwys'.

94 Ibid. 120.

95 Pennant, *Tour in Wales* (1778), 381.

96 *The Llanarmon Urn,* watercolour [jts03036] by Moses Griffith inserted at foot of page 381 in grangerized copy of Pennant's Tour in Wales, MDCCLXXIII (London: 1778), NLW PD09874.

97 Richard Mabey, *Gilbert White: A Biography of the Author of The Natural History of Selborne* (London: Century, 1986), 106.

98 Evans, 'Thomas Pennant (1726–1798)', 405.

99 Patrick O'Brian, *Joseph Banks: A Life* (London: Collins Harvill, 1987), 236.

100 W. T. Parkins, 'Life of the Author', in *Pennant's Tours in Wales*, ed. J. Rhys, 3 vols (Caernarvon: H. Humphreys, 1883), I: xi–xlvi; 'Life of Thomas Pennant, Esq.', *Scots Magazine*, 1 July 1799: 426–32; 'The Late Thomas Pennant, Esq. F.R.S. & F.R.S.U. &c. &c.', *Chester Courant*, 19 February 1799.

101 *Scots Magazine*, 1 January 1776: 4.

102 E.g. 'To be sold [...] the Lands and Barony of Faskally [...] and [...] Lands of Ballyfuirt', *Caledonian Mercury*, 20 July 1776; 'To be sold [...] Lands and Barony of Moness', ibid. 6 August, 13 October 1787.

103 For a useful review of Pennant's critics across the spectrum, see L. F. Powell, 'The Tours of Thomas Pennant', *The Library* 4th series, 19 (1938): 131–54. This text includes contrasting opinions of Pennant from Johnson, Pinkerton and Walpole.

104 *Scots Magazine*, 1 July 1774: 370. For Johnson, see Evans, 'Thomas Pennant (1726–98)', 397.

105 *Scots Magazine*, 1 December 1783: 41.

106 Ibid. 1 June 1808: 444.

107 Ibid.

108 See Appendix XXVII, 'Statement of Facts', in Rhys, ed., *Pennant's Tours in Wales* III: 413–15.

109 Ibid. 415.

110 Farrar Fenton, 'Life of Richard Fenton', in Richard Fenton, *A Historical Tour through Pembrokeshire*, 2nd ed. (Brecon: Davies and Co., 1903), ix–xxxi (xviii). This misunderstanding is repeated by Dilwyn Miles, 'Richard Fenton: Pembrokeshire Historian 1747–1821', *Journal of the Pembrokeshire Historical Society* 6 (1996): 50–66 (61), and is unexplained in Brynley F. Roberts, ODNB Online *s.n.* Fenton, Richard (1747–1821) (accessed 23 May 2016).

111 Fenton, 'Life of Richard Fenton', xxi.

112 For Colt Hoare, see Kenneth Woodbridge, *Landscape and Antiquity* (Oxford: Clarendon Press, 1970); M. W. Thompson, *The Journeys of Richard Colt Hoare through Wales and England 1793–1810* (Gloucester: Alan Sutton, 1983).

.

Chapter 3

HEART OF DARKNESS: THOMAS PENNANT AND ROMAN BRITAIN

Mary-Ann Constantine

*'And this also', said Marlow suddenly, 'has been one of the dark places of the earth [...]
I was thinking of very old times, when the Romans first came here, nineteen hundred years
ago – the other day.*[1]

Thus the thoughtful narrator of Joseph Conrad's great novel, seated cross-legged right aft, leaning against the mizzen mast, with the luminous Thames estuary stretching ahead of him, and the darkness looming behind. Conrad's tightly controlled masterpiece, where every word carries its freight of enigmatic meaning, is perhaps an odd match with Pennant's abundant confusion of narrative voices, where deeper meaning can often appear as much the product of happenstance as intention. And though both are narratives of travel, it is perhaps too glib to compare 'a journey to the remotest part of North Britain, a country almost as little known to its southern brothers as Kamtschatka'[2] with Marlow's journey into the 'place of darkness' in the heart of Africa; a more historically appropriate exotic shadowing of the Scottish tour, as Pat Rogers has shown, would be the exactly contemporaneous exploration being undertaken by Pennant's friends and Royal Society colleagues in the Southern Hemisphere.[3] Nevertheless, like Conrad's novel, Pennant's *Tours*, both in Scotland and Wales, reveal interesting moments when the Roman conquest of Britain pushes its way to the surface of a story about modernity: and both are, undeniably, troubled by their Roman predecessors.

This chapter looks at some examples of Pennant's dealings with the Roman world, both through artefacts encountered on his travels, and through the prism of his Classics-based education, to consider how ideas about Roman Britain may have shaped his perceptions of the multicultural roots of the modern British Isles. His evident fascination with this period of British history, one shared by fellow antiquaries throughout the century, is complicated by contemporary political narratives: while his Unionist inclinations approve the drawing together of the peripheries under a benign and improving larger whole, the echoes of violence and oppression from the distant past unsettle him. Looking at Pennant's treatment of the Romans in both the Scottish and the Welsh *Tours* also helps to articulate certain interesting differences in his approach to the two countries; and offers new ways of understanding the nature of his travel writing.

Since part of my argument relies on close textual analysis, it is worth recalling some of the challenges faced when reading Pennant's prose. As many of the essays in this volume make clear, the *Tours* of both Wales and Scotland are complex, composite texts. Pennant's main narrative technique involves splicing sections of direct observational present tense (what Boswell scathingly referred to as 'curt frittered fragments')[4] with more reflective past-tense narrative based on material often gathered after the event, so that (as we shall see below) the author/narrator's own perspective within the time frame of the text is sometimes difficult to fix. Moreover, as the reader moves through the landscape, encountering artefacts, people, scenes and texts, the vocabularies of different fields of enquiry (antiquarian, aesthetic, scientific, political) generate multiple styles, which are further complicated by the fact that some sections are clearly, others less clearly, *not* Pennant's own. The best-known example of this is the description of Staffa in the Hebridean tour, where he simply hands over to Joseph Banks: within the community of scientific discovery and research, the information takes precedence over the conveyer. In other sections, however, the observations of other authors, or of colleagues and correspondents, are simply subsumed, unacknowledged, into the text. The fact that subsequent editions also tended to acquire accretions of appendices, like barnacles, by a variety of authors, again adds further layers to the reader's experience of the *Tour*. The slipperiness of narrative authority in such writing throws up its own peculiar moments – moments of awkwardness or strain of the kind identified by the French critic Pierre Macherey as symptoms of a broader cultural unease.[5]

Like any educated gentleman of his class and period, Pennant, who attended Wrexham Grammar School, and the Queen's College, Oxford, had an education firmly based on the Classics; Romano-British history via Caesar and Tacitus would have been territory familiar from boyhood, while the close proximity of Downing to Chester must have given the Roman past an early material presence in his imagination. The course of the eighteenth century also saw the literal uncovering of Roman Britain, as scholars identified, excavated and mapped the physical remains of the Roman invasion and military presence: Pennant draws on many of the century's key texts, including

Camden's *Britannia*, revised and translated by Gibson in important editions of 1695 and 1722; William Stukeley's *Itinerarium Curiosum* (1724); Alexander Gordon's *Itinerarium septentrionale* (1726); and John Horsley's epic of antiquarian endeavour, *Britannia Romana* (1734).[6] Rosemary Sweet and others have pointed to the overwhelmingly textual bias of what we now term archaeological scholarship at this period: artefacts, roads and ruined forts were largely interpreted through the prism of the authoritative Classical authors, their material evidence bearing witness to established truths.[7] Pennant's *Tours* offer some interesting examples of the kinds of tensions generated by this approach.

As Paul Evans notes in this volume, Pennant's 1769 *Tour* of Scotland was primarily the account of a natural historian; it is much concerned with wildlife, and makes frequent cross-references to the earlier *British Zoology*, which it amplifies and corrects in the light of new observations. But, like the others, this tour begins in Chester ('the Deva and Devane of Antonine, and the station of the *Legio vicessima victrix*') under the eye of the 'Dea Armigera Minerva, with her bird and her altar on the face of a rock in a small field near the Welsh end of the bridge'.[8] The Roman substratum of British history is never very far away. Indeed, in the detailed and innovative map produced in the wake of his second Scottish tour, Pennant includes a wealth of Roman names alongside their modern counterparts – pulling that Classical substratum back up to the surface, and making it simultaneous with his own journey.[9]

Pennant travelled up through Berwick and into the borders, remarking that 'the entrance into Scotland has a very unpromising look'.[10] Typically, he identifies historical reasons for agricultural neglect: the debatable borderlands, endlessly warring, have lacked the settled society necessary for improvement. Beyond Eyton however it gets better, and the shift in landscape is marked by a self-conscious shift in tenses:

> the wretched cottages, or rather hovels of the country, were vanishing; good comfortable houses arise in their stead; the lands are inclosing [...] the banks are planting: I speak in the present tense; for there is still a mixture of the old negligence left amidst recent improvements, which look like the works of a new colony in a wretched impoverished country.[11]

The idea of a beneficial 'new colony' receives, on the same page, indirect historical (and subliminally military) reinforcement from Pennant's description of the strategic advantages of the Firth of Forth as seen through the eyes of Agricola. The connection between Roman and modern British endeavours, however, becomes most explicit at Taymouth, where, in a lyrical and unusually first-person passage, Pennant celebrates the beautiful and varied landscape around the Breadalbane estate:

It is very difficult to leave the environs of this delightfull place: and, before I go within doors, I must recall to mind the fine widening walks on the south side of the hills, the great beech sixteen feet in girth, the picturesque birch with its long streaming branches, the hermitage, the great cataracts adjacent, and the darksome chasm beneath. I must enjoy over again the view of the fine reach of the Tay, and its union with the broad water of the Lion: I must step down to view the druidical circles of stones, called in the Erse, *Tibberd*; and lastly, I must visit Tay-bridge, and, as far as my pen can contribute, extend the fame of our military countrymen, who, among other works worthy of the Romans, founded this bridge, and left its history inscribed in these terms:

Mirare
viam hanc militarem
Ultra Romanos terminos
M. Passuum CCL. *hac illac*
extensam;
Tesquis et paludibus insultantem
per Montes rupesque patefactam
et indignanti TAVO
ut cernis instratum,
Opus hoc arduum sua solertia
Et decennali militum opera
A Aer. X^{nae} 1733. Posuit G. WADE
Copiarum in SCOTIA Praefectus
Ecce quantum valeant
Regis GEORGII II. Auspicia.[12]

Here the beauty and tranquillity of the well-managed estate (Breadalbane underwent extensive landscaping in the second half of the eighteenth century) are not merely juxtaposed with but linked directly, via Wade's bridge, to Hanoverian military endeavour.[13] This vision of gracious living is made possible by the endeavours of 'our military countrymen', who are 'worthy of the Romans' they so self-consciously emulate (for once, in Pennant's text, there can be no uncertainty about the force of that 'our'). General Wade and his roads make further appearances throughout this tour:[14] at Lochaber another of his impressive bridges is credited with helping to rein in the power of the lawless freebooters of the region, while the even more impressive roads

by rendering the highlands accessible contributed much to their improvement, and were owing to the industry of our soldiery. They were begun in 1723, under the directions of General Wade, who like another Hannibal, forced his way through rocks supposed to have been unconquerable.[15]

And though the general's road building is elsewhere cited as 'one rare example of making the soldiery usefull in times of peace',[16] the language of these endeavours is thoroughly militarized. 'The bogs and moors', writes Pennant, 'had likewise their difficulties to overcome, but all were at length constrained to yield to the perseverance of our troops.'[17] The association between Roman and contemporary military road building was often made explicit in the period, and found expression in various ways. As in the case of Wade's inscription above, Pennant notes that the soldier–roadbuilders wrote themselves quite deliberately into the landscape:

> In some places I observed, that, after the manner of the Romans, they left engraven on the rocks the names of the regiment each party belonged to, who were employed in these works; nor were they less worthy of being immortalized than the Vexillatio's of the Roman legions; for civilization was the consequence of the labors of both.[18]

Meanwhile, in a kind of reverse relationship, the Lanarkshire engineer William Roy used contemporary military maps of Scotland to plot and record Roman sites throughout the 1760s; although his *Military Antiquities* was not published until 1793, Pennant corresponded with him while putting together his own map of the Scottish Highlands during the 1770s.[19]

It is towards the end of the 1769 tour, as he returns south, that Pennant engages with one of the eighteenth century's most important and discussed Roman artefacts, the Antonine Wall, more usually known in the period as Graham's Dyke.[20] He also mentions (and includes an illustration of) the lost Arthur's O'on (Oven), identified by earlier antiquarians as a Roman shrine. For once, it seems, Pennant's 'improving' instincts are quite dampened by antiquarian regret: 'to the mortification of every curious traveller, this matchless edifice is no more; its barbarous owner, a gothic knight, caused it to be demolished, in order to make a mill-dam with the materials'.[21] Visiting Glasgow shortly before seeing remains of the wall itself, Pennant sees the university's growing collection of 'monumental and other stones'. A footnote tells us that engravings of several of these have recently been commissioned, and that 'the Provost of the University did me the honour of presenting me with a set'.[22] As Lawrence Keppie has shown, the University of Glasgow played an important part in collecting and preserving pieces of sculpted stonework from the Wall, acquiring a fine collection of the pieces known as distance slabs. Pennant's compatriot Edward Lhuyd had seen and copied a number of them in 1699, noting that '[t]hey keep these stones at Glascow very carefully in the Library; and the Principal was daily expecting two or three more that had been promised him'.[23] The engravings were produced by the Foulis Academy, a school of art set up in 1753 by the enterprising brothers Robert and Andrew Foulis, who did most of the bookselling and printing in Glasgow at this time. Pennant must therefore have received his set soon after they were first engraved in 1768: they would not be officially published until an expanded edition came out as *Monumenta Romani Imperii* in 1792.[24]

Briefly mentioned in a footnote in 1769, the engravings are given rather more attention in the second Scottish tour of 1772, and one in particular is noted (Fig. 3.1):

> None is more instructive than that engraven in plate III, on which appears a Victory about to crown a Roman horseman, armed with a spear and shield. Beneath him are two Caledonian captives, naked, and bound, with their little daggers, like the modern dirks, by them.[25]

Pennant doesn't say why this image should be 'instructive', or what the lesson to be drawn from it may be, but it is certainly striking, indeed unsettling;

Figure 3.1 Roman distance slab from Summerston, from *Monumenta Romani Imperii* (Glasgow, 1792).

and the small prompt of those 'modern dirks' gives pause. Victorious horse-men and conquered Caledonians are, after all, not merely figures from the deep past, but from recent British history; and though the description itself gives relatively little away, the cadences evoke a certain solemnity, if not pity.[26]

Pennant in Scotland is undeniably more Roman than native: he is the rider on horseback, he describes, controls, and classifies. His belief that the Jacobite defeat at Culloden was the saving of 'North Britain' is stated clearly enough at several points, as is his faith in the potential for rapid economic improve-ment under an enlightened centralized British government. Encounters with natives, however, whether historically distant Caledonians or people encoun-tered on his journey (and, as Nigel Leask has pointed out, the two, inevita-bly, are sometimes conflated) often produce moments where the text seems to pull in different directions.[27] Sometimes this is a direct result of the par-ticular discourse involved. Pennant's observational present tense, for example, employs a somewhat telegraphed log-book style, which leaves plenty of room for interpretation:

> A boat filled with women and children crosses over from Jura, to collect their daily wretched fare, limpets and perriwinkles. Observe the black guillemots in little flocks, very wild and much in motion.[28]

The restraint of the observational mode feels surprisingly modern. No lesson is being drawn here, no explicit parallel; yet the women and children, like the guillemots, are hunting for food. (The guillemots, it appears, are probably doing rather better, nutritionally speaking, than the humans.) Though the sci-entific gaze is arguably here infused with sympathy by the word 'wretched', one might also read this clipped presentation of misery as somewhat detached; or be troubled by the proximity of the scavenging Jurans and the words 'very wild' – as if the correct classification for Jurans is, in fact, with seabirds. But the text itself gives nothing much away.

A more expanded episode later in the journey, however, gives more depth to Pennant's response. The party arrive at the island of Canna on a sunny June evening and anchor in the 'snug' harbour. It appears, in Pennant's words, 'pleasing to humanity; verdant, and covered with hundreds of cattle', giving 'a full idea of plenty':

> but a short conversation with the natives soon dispelled this agreeable error: they were at this very time in such want, that numbers for a long time had neither bread nor meal for their poor babes: fish and milk was their whole subsistence at this time: the first was a precarious relief, for, besides the uncertainty of their success, to add to their distress, their stock of fish-hooks was almost exhausted;

and to ours, that it was not in our power to supply them. The rubbans, and other trifles I had brought would have been insults to people in distress. I lamented that my money had been so uselessly laid out; for a few dozens of fish-hooks, or a few pecks of meal, would have made them happy.[29]

This ghastly colonial moment (and to his credit Pennant appears to recognize it as such) is a Hebridean echo of the arrival of Cook's expedition in Tahiti three years earlier, where 'beads and small presents' had played their part in an elaborate welcome.[30] Pennant is sharp enough to see how inappropriate such gifts are here, and compassionate enough to grieve for that missed opportunity to help. Unlike many later tourists, focused on the aesthetics of landscape and the experience of the traveller, he gets beyond the surface, beyond the beauty and the illusion of plenty, to some understanding of the difficult conditions of the inhabitants.

A constant theme of the Scottish *Tours* is the nature of the responsibility of government at all levels. Pennant can be very critical of local communities who mismanage their resources, but he is equally scathing of wider systemic failures to prevent this kind of distress. The responsibility and to some extent the blame for the situations are, like his attitude to the people themselves, partly shaped by the different modes of writing employed within the text – and again, this throws up some interesting moments. Becalmed for several days in the straits of Jura, Pennant informs his readers that he decided to write up a history of the Hebrides. This is then inserted into his narrative of the tour, enacting the becalming by stopping the reader's forward progress: we too are obliged to stop, and take stock of the past. The section (as he acknowledges) draws on a chapter from the recently and posthumously published *Critical Dissertations* of the Revd Dr John Macpherson of Sleat on Skye, and the 'voice' cataloguing a history of constant and rather violent political flux in the islands is for the most part unobtrusively historical.[31] At the very end, however, as the narrative approaches the present, the tone changes. The heads of the clans in recent times, it appears, have been unwisely courted and flattered as desirable allies, rather than 'treated as bad subjects':

> Two recent rebellions gave legislature a late experience of the folly of permitting the feudal system to exist in any part of its dominions. The act of 1748 at once deprived the chieftains of all power of injuring the public by their commotions. Many of these *reguli* second this effort of legislature, and neglect no opportunity of rendering themselves hateful to their unhappy vassals, the former instruments of ambition. The *Halcyon* days are near at hand: oppression will beget depopulation; and depopulation will give us dear-bought tranquility.[32]

This is the voice of the victorious Roman horseman: the civilizer, the extender of dominions. It is a voice which believes in strong governance, and in the inherent rottenness of a feudal system which its abstract but powerful 'legislature' seeks to replace. In the last two sentences, however, something curious happens: we are told that the corrupt clan chieftains are helping the cause of 'legislature' by being hateful to their own people. Whether Pennant intended it or not, it is suddenly quite hard not to put the corrupt clan chiefs and 'legislature' in the same category of 'oppression' – 'oppression will beget depopulation; and depopulation will give us dear-bought tranquility.' And who, exactly, is this 'us'? If 'us' is the British people as a great unified whole, the forces of legislature embodied, then the reader might wonder why the absence of a few rabble-rousing Highlanders should be perceived as 'dear-bought'? Yet the scenario of the Highlands and Islands emptying is clearly presented as something that should shock us. The ironic, indeed bitter, tone of the last sentence sits oddly with the rest.

This may well be, as Nigel Leask has suggested, an expression of Pennant's fear that depopulation will weaken Britain's military reservoir in the Highlands.[33] But it is also the result, once again, of a submerged voice, a textual echo, interfering with the general message. The voice in this instance is that of Calgacus, the fiery Caledonian chieftain of Tacitus's *Agricola* who, before the battle of Mons Graupius (AD 83 or 84), is imagined addressing an alliance of Caledonian tribes. Tacitus, ventriloquizing nicely, gives him a famous and blisteringly anti-Roman speech:

> Robbers of the world, now that earth fails their all-devastating hands, they probe even the sea: if the enemy have wealth, they have greed; if he be poor, they are ambitious; East nor West has glutted them; alone of mankind they covet with the same passion want as much as wealth. To plunder, butcher, steal, these things they misname empire; they make a desolation and they call it peace.[34]

'Depopulation' and 'dear-bought tranquility' are thus, in Pennant's account, submerged Tacitean terms for a scenario produced by the ruthless ambitions of Empire; it is hardly surprising therefore that they chafe against the predominant message of the section, which is trying to persuade its readers that latter-day Caledonians are not fit to rule themselves. The ghost of Calgacus – though barely perceptible – is a disruptive presence here.

In its own subtle way, the passage reflects a deep duality in eighteenth-century antiquarian responses to the idea of *Romanitas*, as well as highlighting the multiple and sometimes conflicting political positions of the *Agricola* itself.[35] As Richard Hingley has shown, the broader context of debates about unification ensured that Scottish writers on Roman antiquity almost inevitably used either the Antonine or Hadrian's Wall to 'explore the differing identities of contemporary populations'.[36] These competing categories of identity, with their reversible polarities, spoke directly to contemporary concerns about governance, the global expansion of British interests, and the nature of civil society in the eighteenth century. The Wall could be conceived as a line defining Roman civility against Caledonian barbarity, or, conversely (and largely thanks to 'Calgacus') a defiant demarcation protecting the virtues of Caledonian simplicity and bravery against Roman luxury and effeteness. The dichotomy frequently split along Scottish versus English lines; but *Romanitas* could also be claimed as part of a Lowland Scots inheritance, and set against a rejected (or romanticized) Highland other. Traces of all these positions, possibly absorbed from earlier writers like Alexander Gordon, can be found in Pennant's writings; much like the Classical sources themselves, the multi-voiced nature of his writing allows for competing interpretations.

Calgacus also makes an appearance in *A Tour in Wales*, which came out in three parts between 1778 and 1783. The Advertisement to the first volume, titled *A Tour in Wales 1773*, assures its readers from the outset that, though published some five years later:

> [T]hese home travels are the first part of an account of my own country; and were actually performed in the year mentioned in the title page. The world justly loves the reality; therefore, this is mentioned to satisfy the public, that they are not formed out of tours undertaken at different periods.[37]

This assurance of temporal veracity disappears in later editions of the text, as Pennant, true to form, adds more and more information from further excursions and from his network of Welsh correspondents. But it is worth noting how swiftly that first exploratory Welsh tour took place after his return from Scotland. Indeed, the two countries are closely linked, and jointly located in a shared and defiant past, in the exuberant (and rather Virgilian) opening lines of the *Tour* itself:

> I now speak of my native country, celebrated in our earliest history for its valour and tenacity of its liberty; for the stand it made against the *Romans*; for its slaughter of the legions; and for the subjection of the nation by *Agricola*, who did not dare to attempt his Caledonian expedition, and leave behind him unconquered so tremendous an enemy.[38]

The Welsh, then, are also part of this Tacitean story, and have their own history of resistance to Agricola; the entire Caledonian campaign is made dependent on first conquering this 'tremendous enemy'. As in the Scottish *Tours*, that troubling history of conquest and absorption into *Romanitas* also lurks beneath the surface at certain moments in the text.

Some fifty pages into the first tour, in the environs of Flint, the discovery of a number of metal Roman antiquities leads Pennant to a discussion of the area's ancient lead-mining operations. The nature and processes of these workings absorb Pennant (though, one fears, probably not all of his readers) for several dense and wide-ranging pages, in which he draws heavily on Classical authors to discuss the nature of mining in Britain before and after the Roman conquest:

> Previous to the settlement of the Romans in Britain, Strabo speaks so slightly of our articles of commerce, as to say, they were not worth the expence of one legion and a few horse. He died in the year 25, before our country was scarcely known, except by the attempt of Caesar. But the trade, both in his days, and those of that great geographer, was carried on merely by exchange. The Britons worked their own mines of tin and lead; and in their room received from the foreign merchants, earthen-ware, salt, and works of brass.
>
> In a small time after the Romans had carried their arms through our islands, they began to apply with vigor to the workings of the mines.[39]

These early, low-key attempts of the native Britons at exploiting their natural resources are energized and professionalized by Roman 'vigor', in a spirit of improvement entirely in keeping with Pennant's own position as an eighteenth-century landowner keen to see all the resources of the united British Isles used as productively as possible. The problem comes some pages later, when he turns to consider the labour itself, and it is here that Calgacus makes his entrance:

> The miners, in the earlier times of the Romans in Britain, seem to have been the subdued natives. Galgacus encourages his soldiers to conquer or die, by laying before them the dreadful consequences of a defeat: *Tributa et Metalla, et caeterae serventium poenae.* 'Tributes and mines, and all the dire penalties of slavery'. Agricola himself verifies the prophetic spirit of our brave chieftain, by calling our mines the reward of victory. These were to be worked, not by the conquerors, but by condemned criminals, by slaves,[z] and Britons newly subjugated. It is probable, that when the island was entirely settled, this badge of slavery was taken away and the miners were, as before the arrival of the Romans, voluntary laborers.[40]
>
> [z] *Diodorus Siculus*, lib.v.c.2. gives a melancholy account of these slaves; whose state can only by paralleled by the poor *Indians* in the mines of *Potosi*.

The paragraph itself closes down very quickly on the troubling thoughts it has raised: even ancient Britons never shall be slaves, but rather 'voluntary laborers' towards a greater productive good. That footnote to Diodorus Siculus, though, is nagging, insistent, evoking as it does a harsh description of the conditions of slaves in the silver mines of Spain under Roman rule. In the *Bibliotheca Historia*, written between 30 and 60 BC, Diodorus describes how

> the slaves who are engaged in the working of them produce for their masters revenues in sums defying belief, but they themselves wear out their bodies both by day and by night in the diggings under the earth, dying in large numbers because of the exceptional hardships they endure. For no respite or pause is granted them in their labours, but compelled beneath blows of the overseers to endure the severity of their plight, they throw away their lives in this wretched manner.[41]

The modern reference to Potosi, a silver mountain in Bolivia and a major source of Spanish colonial wealth, also summons up the image of subjugated native labourers worked relentlessly to death; the cruelty of what James Thomson called 'sad Potosi's mines' was an established trope in eighteenth-century writing.[42] A recent study by Jane Aaron of nineteenth-century novels set in Wales suggests that plots involving mines and miners might be read as a form of 'Gothic' – that they act, in effect, as a kind of troubled narrative subconscious.[43] Pennant's indirectly expressed anxieties about the Roman mines perhaps do something similar: like Marlow's oblique but ineradicable description of black labourers crawling off to die in a pit ('it was just a hole'), this text reveals glimpses of the human cost of the infrastructure of empire.

This is not to suggest that every encounter with Roman remains is fraught with anxiety. Much of the time the artefacts and buildings are simply the objects of a profound interest and admiration – from the impressive collection at Senhouse in Cumbria or the detailed accounts of Roman Chester, to the pleasing discovery that the inhabitants of Perth are still proud to recall Roman enthusiasm for their river Tay (*'Ecce Tiberim!'*).[44] There are, however, enough examples like those already discussed above to suggest a persistent underlying tension between the centre and the peripheries: far from providing, as Paul Smethurst has suggested, 'evidence of a shared culture and history that might provide the foundation and frame for a reinvented, cohesive topography of Britain',[45] Pennant's explorations into British history tend rather to stir up trouble. On Anglesey – ancient Mona – he enters 'classical ground' and is reverently entranced by 'the pious seats of the antient Druids; the sacred groves, the altars, and monumental stones'; a vision quickly destroyed by the violent irruption of the invading armies of Suetonius 'who put an end in this island to the Druid reign'.[46]

One final, striking, encounter with the Caledonian past plays out some of these tensions in rather an unexpected way. At the end of Volume I of the 1772 *Tour of Scotland*, Pennant finishes his sailing tour in Ard-maddie. He retires to his chamber, unpacks and reflects, with enormous gratitude for the help he has received, on 'this voyage of amusement, successful and satisfactory in every part, unless where embittered with reflections on the sufferings of my fellow-creatures'.[47] He falls asleep, and begins to dream of the feudal past, 'of heroes immortalized in the verse of OSSIAN', and of 'an antient warrior', who addresses him thus:

> STRANGER, Thy purpose is not unknown to me; I have attended thee (invisible) in all thy voyage; have sympathised with thee in the rising tear at the misery of my once-loved country; and sighs, such as a spirit can emit, have been faithful echoes to those of thy corporeal frame.[48]

The vision, then, for some seven or eight pages, delivers a gloriously florid account of the clan system in the good old days, when warriors were heroic, and vassals were loyal, and tables groaned under the weight of the meat and the mead; when rights and duties were expressed through reciprocal bonds of love, friendship, respect, loyalty and protection.

It is, as many others have noted, a persuasive technique: Pennant's dream vision allows him to voice a deeper, more sympathetic – albeit idealized – insight into the culture of the clans. It is perhaps the closest he comes to going native: his Mr Kurtz moment. And, exactly as with the earlier example discussed above, it is at precisely the same point of transition, from the glorious past into the troubled present, that the chieftain's voice breaks, cracks, and another voice starts to take over:

> My progeny for a while supported the *great and wild magnificence of the feudal reign.* Their distance from court unfortunately prevented them from knowing they had a superior; and their ideas of loyalty were regulated only by the respect or attention paid to their fancied independency. Their vassals were happy or miserable, according to the disposition of the little monarch of the time. Two centuries, from my days, had elapsed, before their greatness knew its final period. *The shackles of the feudal government* were at length struck off; and possible happiness was announced to the meanest vassal [...] The mighty Chieftains, the brave and disinterested heroes of old times, *by a most violent and surprizing transformation*, at once sunk into rapacious landlords [my italics].[49]

In a few brief sentences the feudal system has moved from magnificence to oppression; a 'violent and surprizing' transformation indeed. In this awkward paragraph, one feels, Pennant is trying to scramble back onto his horse.

The ghost then accuses his own degenerate descendants of having abandoned Scotland, and of seeking advancement in foreign climes; he urges them to return home and live with the people in their land, educating them, teaching them to fish and farm with the new methods. Then – and again the pronouns become exceptionally difficult to pin down – he asks his listeners (the degenerate descendants, or a wider British readership?) to ensure that the Scottish people become good and loyal citizens to the state. Train them, he says, for *your* navy, for *your* army, to fight against the real foreign enemies ('Have not thousands in the late war proved their sincerity? Have not thousands expiated with their blood the folly of rebellion, and the crimes of their parents?'). And if you continue to neglect them, he adds, do not be surprised if they join with 'our natural enemies' ('How dreadful will be the once-existent folly of Jacobitism, transformed into the accursed spirit of political libertinism!').

The ghost concludes with a thundering attack on those who can sit down to dinner while their compatriots are dying of hunger: can you, he asks, 'feel a momentary remorse for deaths occasioned by *ye, ye* thoughtless deserters of your people'?. But *whose* people are they now, and who exactly is doing the deserting? If these people who were so recently on the wrong side of the notional wall, beyond the Pale, are now enclosed and incorporated, then where should responsibility for their suffering lie – with the old regime which caused it, or the new regime which takes it on? Once again, the devil is in the pronouns ('you/your', 'we/our') which seem simultaneously addressed by two voices to two quite different audiences: the imaginary chieftain to his Scottish descendants, and Pennant to his wider British readership. Likewise, it is hard not to read the Chieftain's very last words as similarly double-voiced, since they close Volume I and the second volume did not appear for another two years:

> With all my failings, I exult in innocence of such crimes; and felicitate myself on my aerial state, capable of withdrawing from the sight of miseries I cannot alleviate, and of oppressions I cannot prevent.[50]

The traveller, like the vision, can at least withdraw.

One of the most interesting and compelling aspects of the 1772 Scottish *Tour* is its clear signs of distress with 'miseries I cannot alleviate, and oppressions I cannot prevent', and the ways this distress manifests itself in fault lines in the text. Roman horseman notwithstanding, it may be that what Nigel Leask has termed the 'understated Welshness'[51] of this Flintshire landowner contributes to his sympathetic and thoughtful reaction to rural poverty; paradoxically more so in Scotland, it has been suggested, than in his native land.[52] During the decade after the French Revolution, when rebellion was potentially closer to home, Pennant became far less liberal in outlook, and his concerns for the poor (who were rioting over corn prices in nearby Wrexham and Ruthin) appear far more self-interested, a matter of preventing rebellion on his own doorstep. Yet, even if his autobiographical *Literary Life* of 1793 is more than a touch smug in its recollections, there is no doubt that the tour of Scotland had direct material and intellectual consequences for the country it described:

> IN this tour, as in all the following, I laboured earnestly to conciliate the affections of the two nations, so wickedly and studiously set at variance by evil-designing people. I received several very flattering letters on the occasion […] My success was equal to my hopes; I pointed out everything I thought would be of service to the country; it was rouzed to look into its advantages; societies have been formed for the improvement of the fisheries, and for the founding of towns in proper places.[53]

The traveller's 'gaze', so often figured as detached and ethereal (if not, indeed, downright callous) can, providing the traveller is well connected enough, actually make things happen on the ground.

Conclusion

The printed itineraries at the back of Pennant's *Tours* diligently (if not always accurately) mark up all the Roman names of places he passes through on his journeys;[54] it is striking how closely, in eighteenth-century scholarship, Roman and modern itineraries are intertwined. The antiquarian walkers and riders of real or perceived Roman roads mapped out the deep past through their own journeys – and their own published *itineraria* – to summon up a world of long marches, milestones, ditches and camps. As Rosemary Sweet observes, for most of the century Roman Britain was essentially 'a military concept' and as a result very few historians attempted to discover traces of domestic Romano-British life: farms and villas and towns were less imaginatively present than routes and ramparts, lines of attack or defence.[55] The Industrial Revolution, with its rapid improvement of travel infrastructure, not only made such antiquarian tours easier, but also played its part in uncovering (and not infrequently destroying) lost Roman roads. And, in Scotland at least, the shadow of much more recent military history meant that for the Classically educated eighteenth-century traveller curious about the past, Roman soldiers were ghostly and inevitable travelling companions.

One minor Roman episode, however, has a more domestic flavour. In the opening pages of the 1769 *Tour*, Pennant travels through the spa town of Buxton and comments effusively on its 'celebrated warm bath':

> with joy and gratitude I this moment reflect on the efficacious qualities of the waters; I recollect with rapture the return of spirits, the flight of pain, and the re-animation of my long, long crippled rheumatic limbs.[56]

A footnote reminds us that Romans, who were 'remarkably fond of warm baths', also sought out the Buxton springs, but then this hymn to 'Hygeia' is immediately followed by the more sobering reflection that 'what Providence designed for the general good, should be rendered only a partial one, and denied to all, except the opulent', whose ailments, caused by overindulgence, are so often self-inflicted. It is to be hoped, continues Pennant, that the successor to the late Duke of Devonshire will carry forward his project of enclosing a further nine springs in the grounds of the Hall, and making them available,

> not solely to those whom misused wealth hath rendered invalids, but to the poor cripple, whom honest labour hath made a burden to himself and his country, and to the soldier and sailor who by hard service have lost the use of those very limbs which were once active in our defence.[57]

Even warm baths are not as comforting a subject as they might be, leading as they do to thoughts of the excluded, and the military maimed. And once again a Tacitean undertow brings its own ripple of disturbance here, as the Roman author famously equated the trappings of civilization with the act of military conquest:

> so the Britons were gradually led astray by the allurements of vice – porticos, baths, elegant banquets. To the inexperienced all that is called 'civilization' (humanitas) when in fact it was only a feature of their enslavement.[58]

'Porticos, baths and elegant banquets' could scarcely be bettered as a thumbnail portrait of eighteenth-century polite society at its most desirably Roman; and yet every schoolboy will have understood the phrase to be double-edged. Far from being merely a cosy marker of belonging to a 'gentlemen's club', knowledge of Latin and Greek authors gave educated writers – and perhaps especially travelling antiquarians – an extraordinarily flexible set of positions from which to contemplate the history, economics and power relations of the British Isles, past and present.[59] It is not surprising then, that like the Greek and Roman authors he cites or subconsciously evokes, Pennant sometimes

finds himself, on his journeys at the edges of the new Britannia, ventriloquizing expressions of unease or frustration with the centralizing powers he
represents.

Notes

1 Joseph Conrad, *Heart of Darkness* (Ware: Wordsworth Classics, 1999), 33.
2 Thomas Pennant, *The Literary Life of the Late Thomas Pennant, Esq., by Himself* (London, 1793), 11.
3 Pat Rogers, *Johnson and Boswell: The Transit of Caledonia* (Oxford: Clarendon Press, 1995).
4 James Boswell, *The Life of Samuel Johnson LLD*, 2 vols (London, 1791) I: 216.
5 Pierre Macherey, *Pour une Théorie de la Production Littéraire* (Paris, 1966), argues that 'ideology', in literary texts, reveals itself through gaps and fissures, through what is *not* said;
  and so the critic's task, as Terry Eagleton explains it, is 'to produce a new discourse
  which "makes speak" the text's silences'; see Terry Eagleton, 'Pierre Macherey and the
  Theory of Literary Production', *Minnesota Review*, new series 5 (Fall 1975): 134–44.
6 Richard Hingley, *The Recovery of Roman Britain, 1586–1906: A Colony so Fertile*
  (Oxford: Oxford University Press, 2008); Rosemary Sweet, *Antiquaries: The Discovery of
  the Past in Eighteenth-Century Britain* (London: Hambledon and London, 2004); H. Toller,
  'Cambrian Antiquity – The Romanists', in *Archaeology of the Roman Empire*, ed. N. J.
  Higham (Oxford: Archaeopress, 2001), 123–30.
7 Sweet, *Antiquaries*, 177; Hingley, *Recovery of Roman Britain*, 4–7.
8 Thomas Pennant, *A Tour in Scotland 1769* (Chester, 1771), 2. The shrine in Edgar's
  Field is still in situ, now a Grade I listed building. The site is revisited more fully in the
  Chester section of the *Tour in Wales*.
9 Gwyn Walters, 'Thomas Pennant's Map of Scotland, 1777: A Study in Sources, and an
  Introduction to George Paton's Role in the History of Scottish Cartography', *Imago Mundi:
  The International Journal for the History of Cartography* 28, issue 1 (1976): 121–28. This map,
  and Pennant's routes through Scotland, can be accessed at http://curioustravellers.ac.uk/
  map/.
10 Pennant, *Tour in Scotland 1769*, 40.
11 Ibid. 40–41.
12 Ibid. 80–81. The Latin reads: 'Admire this military road, carried on both sides [of the
  river] for 250 miles beyond the Roman bounds, defying moors and marshes, opened
  through rocks and mountains, and laid, as thou seest, across the indignant Tay. This
  arduous undertaking, through his own skill and 10 years' labour of his soldiers, was
  completed in the year 1733 of the Christian era by G. Wade, commander of the forces
  in Scotland. See how beneficent is the royal favour of George the Second.'
13 Note that Breadalbane is one of the four improving landowners cited with approba-
  tion by the ghost of the Caledonian chieftain at the end of the first volume of the 1772
  Scottish *Tour* (see below).
14 The Irish-born British army officer George Wade served in numerous campaigns,
  including the 1715 and 1745 Jacobite Rebellions, and spent many years surveying and
  building military roads in Scotland.
15 Pennant, *Tour in Scotland 1769*, 184.
16 Ibid. 170.
17 Ibid. 185.
18 Ibid.

19 Walters, 'Thomas Pennant's Map of Scotland', 121–28.

20 Lawrence Keppie, *The Antiquarian Rediscovery of the Antonine Wall* (Edinburgh: Society of Antiquaries of Scotland, 2012); idem, *Roman Distance Slabs from the Antonine Wall: A Brief Guide* (Glasgow: Hunterian Museum, 1979).

21 Pennant, *Tour in Scotland 1769*, 212. Pennant here echoes the furious controversy caused by the destruction of this monument amongst the Society of Antiquaries in the 1740s. The vilified landowner was the subject of a satirical drawing by Stukeley. Keppie, *Antiquarian Rediscovery*, 88–89.

22 Pennant, *Tour in Scotland 1769*, 202.

23 Keppie, *Antiquarian Rediscovery*, 55.

24 The *Monumenta* itself is now very rare, though a copy given to William Hunter can be seen at the Hunterian Museum in Glasgow. For the Foulis Academy, see http://special. lib.gla.ac.uk/exhibns/foulis/pupils.htm (accessed 30 May 2015).

25 Thomas Pennant, *A Tour in Scotland, and Voyage to the Hebrides, MDCCLXXII*, 2 vols (Chester, 1774; London, 1776), I: 158. The stone, from Summerston Farm, is described by Keppie, *Roman Distance Slabs*, 14.

26 Lawrence Keppie's recent description of a similar piece in the series, the more detailed Bridgeness slab, is also unexpectedly evocative; see *Roman Distance Slabs*, 5: 'A cavalry-man, representative of the all-conquering Roman forces, rides roughshod over four naked and disarmed native warriors; it is an unequal struggle.'

27 Nigel Leask, 'Thomas Pennant's Scottish Tours: Travel, Knowledge Networks and National Description' (unpublished lecture delivered in Berkeley, 10 November 2014. I'm grateful to the author for a sight of this paper).

28 Pennant, *Tour in Scotland [...] 1772*, I: 242–43.

29 Ibid. 311.

30 Joseph Dalton Hooker (ed), *The Journal of the Right Hon. Sir Joseph Banks* (Cambridge: Cambridge University Press, 2011), 74.

31 John Macpherson, *Critical dissertations on the origin, antiquities, language, government, manners, and religion, of the antient Caledonians, their posterity the Picts, and the British and Irish Scots* (London, 1768). The work was published after his death by his son, and dedicated to Charles Greville; Johnson and Boswell disapproved of it heartily.

32 Pennant, *Tour in Scotland [...] 1772*, I: 242.

33 Leask, 'Thomas Pennant's Scottish Tours'.

34 Tacitus, *Agricola*, trans. M. Hutton, revised by R. M. Ogilvie (Cambridge, Mass. and London: Harvard University Press, 1970), chapter 30, 81.

35 The multiple possibilities of this text for a later British colonial audience are nicely explored by Mark Bradley, 'Tacitus' *Agricola* and the Conquest of Britain', in *Classics and Imperialism in the British Empire*, ed. Mark Bradley (Oxford: Oxford University Press, 2010), 123–57. Howard D. Weinbrot notes some other eighteenth-century literary references to Calgacus in *Britannia's Issue: The Rise of British Literature from Dryden to Ossian* (Cambridge: Cambridge University Press, 1993), 279.

36 Hingley, *Recovery of Roman Britain*, 121, 128, notes that Sir John Clerk of Penicuik adopted a self-consciously Roman identity, taking the name 'Agricola' in William Stukeley's Society of Roman Knights; Alexander Gordon, on the other hand, took the name 'Galgacus' on his election to the same society in 1724. For Stukeley's 'Equites Romanii' and their involvement with the archaeology of Roman Britain, see Philip Ayres, *Classical Culture and the Idea of Rome in Eighteenth-Century England* (Cambridge:

Cambridge University Press, 1997), 91–97. Walter Scott's *The Antiquary* (1816) would poke fun at a century's worth of scholarly 'Agricolamania'.

37 Pennant, *A Tour in Wales 1773* (London, 1778), 'Advertisement'; the later editions omit all but the first clause.

38 Ibid. 1.

39 Ibid. 51.

40 Ibid. 55.

41 Diodorus Siculus, *Library of History*, trans. C. H. Oldfather (Cambridge, Mass. and London: Harvard University Press, 1939), Book V, section 38, 195–96.

42 James Thomson, *The Seasons*: 'Summer', lines 870–72: 'Ah, what avail their fateful treasures, hid / Deep in the bowels of the pitying earth / Golconda's gems, or sad Potosi's mines / Where dwelt the gentlest children of the sun?'

43 Jane Aaron, *Welsh Gothic* (Cardiff: University of Wales Press, 2013), especially the section on 'Coalfield Gothic', 98–107.

44 Pennant, *Tour in Scotland 1769*, 69.

45 Paul Smethurst, *Travel Writing and the Natural World: 1768–1840* (Basingstoke: Palgrave Macmillan, 2012), 110.

46 Pennant, *Tour in Wales 1773*, II: 229, 231.

47 Idem, *Tour in Scotland [...] 1772*, I: 420.

48 Ibid. 422.

49 Ibid. 424.

50 Ibid. 428.

51 Leask, 'Thomas Pennant's Scottish Tours'.

52 John Barrell, *Edward of Pugh of Ruthin: A Native Artist* (Cardiff: University of Wales Press, 2013), 175.

53 Pennant, *Literary Life*, 15.

54 Like many scholars of his time, Pennant (with some misgivings, apparently) made use of the forged Roman *itinerarium* attributed to 'Richard of Cirencester' but which was in fact the work of an eighteenth-century scholar based in Copenhagen. See Sweet, *Antiquaries*, 175–78.

55 Ibid. 181.

56 Pennant, *Tour in Scotland 1769*, 4.

57 Ibid.

58 Bradley, 'Tacitus' *Agricola*', 125. Cf. Edmund Burke's comment of 1760, that Agricola 'subdued the Britains by civilizing them; and made them exchange a savage liberty for a polite and easy subjection' (cited in Hingley, *Recovery of Roman Britain*, 152).

59 Cf. the comment by Jonathan Sachs, *Romantic Antiquity: Rome in the British Imagination, 1789–1832* (Oxford: Oxford University Press, 2010), 33: 'it is clear that the Roman past, imperial or republican, was crucial in C18th for articulating a set of coherent yet flexible set of models with which one could both attack or defend various models of political power and competing understandings of an emergent national identity'.

Chapter 4

CONSTRUCTING IDENTITIES IN THE EIGHTEENTH CENTURY: THOMAS PENNANT AND THE EARLY MEDIEVAL SCULPTURE OF SCOTLAND AND ENGLAND

Jane Hawkes

Introduction: Travel, Taste and the Antique

One of the many phenomena marking the 'long eighteenth century' in Britain was an increased interest and participation in travel with the concomitant activities of acquiring, describing and illustrating art and antiquities, and the production of publications catering for and encouraging these trends.[1] To some extent, these interests were reflected in (and indeed energized by) the works of Johann Joachim Winckelmann, whose *Thoughts on the Imitation of Greek Works*, published in German in 1755,[2] discussed the ideal of 'noble simplicity and quiet grandeur',[3] an ideal that he identified with the Classical, and which he considered it necessary to imitate. The work was an instant success, being reprinted several times and widely translated. In Britain, it enjoyed popularity in artistic circles, with Henry Fuseli's translation being published in 1765, and a second edition appearing in 1767. For Winckelmann himself, the book allowed him to continue his studies in Rome, where he arrived in November 1755, and proceeded to devote himself to the study of Roman antiquities: his method of careful observation enabled him to identify Roman copies of Greek art, something that was less well known and therefore unusual at the time, and led to Roman culture being elevated to the status of the ultimate achievement of Antiquity. As Fuseli's English translation put it: 'There is but one way for the moderns to become great, and perhaps unequalled; I mean, by imitating the ancients.'[4] Imitation, of course, did not mean copying, for this was deemed to be 'the slavish crawling of the hand and eyes, after a certain model';[5] rather, if handled with reason, it could result in that which was imitated assuming another nature and becoming something in its own right: 'reasonable imitation takes just the hint, in order to work by itself',[6] for whatever 'imitation produced, differs from the first idea, as the blossoms of a transplanted tree differ from those that sprung in its native soil'.[7] The mimetic character of art that imitates but does not simply copy was, for Winckelmann, central to any interpretation of Classical idealism,[8] and it is generally understood to mark the early stage of the transformation of Taste in the eighteenth century.

Much has been written on this subject,[9] but here it is worth noting that Winckelmann articulated attitudes generally underpinning perceptions of 'the ancients' at the time in terms of what was considered to be Taste among certain circles of society: specifically, among those interested in art and antiquity, and those most involved in travel and its associated activities, which of course included Thomas Pennant. As has long been recognized, the value of Taste, travel and interest in 'the ancients' was deemed to lie in the exposure it provided to the cultural legacy of Classical antiquity, and for those making the Grand Tour, Italy was one of the most important destinations.[10] As Johnson put it in a letter to Boswell: 'a man who has not been in Italy, is always conscious of an inferiority from his not having seen what is expected a man should see'.[11] As a result, a thriving art market developed, with the tourists themselves sometimes engaging in their own artwork, or travelling with their own artists.

While such activities ensured widespread familiarity with the art and monuments of antiquity, the field of Classical archaeology also began to emerge in a way that might be recognized today – with, most famously, the recovery of the sites at Herculaneum and Pompeii – and the sale, or straightforward expropriation, of the finds. But such excavations were not limited to the sites around Naples. James 'Athenian' Stuart's stay in Rome coincided in 1748 with the rediscovery of the obelisk that had once stood as the gnomen of Augustus' Horologium in the Campus Martius – off the Piazza del Parlamento, where Sixtus V had attempted to repair and raise it after its initial discovery in 1502. It was finally restored and raised by Pius VI between 1789 and 1792. Stuart, who had been earning money painting souvenir fans and easel paintings while acting as a guide to the local 'curiosities and antiquities', was appointed by Cardinal Valenti to record the monument, measuring and illustrating it. The results were published in 1750 as an antiquarian treatise written in Latin and Italian.[12] Filled with citations from Pliny and Strabo, it also included a 100-page appendix by Angelo Maria Bandini of the Vatican library, a figure well known to early medievalists for his work on the Anglo-Saxon Codex Amiantinus of c.700 after he relocated to the Laurenziana in Florence in 1756.[13] His and Stuart's publication on the horologium obelisk attracted important aristocratic patronage which ensured its circulation amongst those who considered themselves to be 'cultivator[s] of the fine arts',[14] and even today it is considered 'a landmark in the history of archaeology'.[15]

It is against this background that, having undertaken a short (five-month) Grand Tour in 1765 in the interests of 'empirical discoveries' and philosophical learning,[16] Thomas Pennant drew attention to vernacular traditions of stone carving in antiquity during his travels through Scotland in 1769 and 1772: to the early medieval (early Christian) sculptures of Scotland, recording his encounters with these monuments in such a way that he was able not only to present them to his readers but also, I suggest, to articulate Scottish ('national') identity in terms of 'ancient' Scottish history. In part his account seems to have been undertaken in the light of earlier antiquarian travels, a tradition established most memorably by Camden in his *Britannia* of 1586; this had given rise to a number of chorographical accounts of the various regions of Britain and Ireland, and in 1722 had been revised and expanded into two large folio volumes by Edmund Gibson in keeping with a current attitude of 'blatant Hanoverian loyalism'.[17] Informing much of this literature was a perceived need to establish the 'rightful place' of Britain within both the world of Roman antiquity and that of an international scholarship focused on the Classical past.[18] In addition to any antipathy Pennant may have felt towards the 'fashionable' motives inspiring travel in Europe at the time, it was this literary tradition and its objects that informed his account of his tours, and the classicizing language he used to present the sculptures of the vernacular antiquity that he encountered in Scotland.

Travel, Taste and the Antique in Britain

Pennant's accounts of the early monuments mark an innovative approach to the material, and one that was to have considerable impact. He was, of course, not the first to travel through the north of England and into Scotland as the increased interest and participation in travel across the European Continent had also come to encompass Britain itself.[19] Celia Fiennes, for instance, had undertaken a series of tours between 1685 and 1703, seeking to cure what was already (for her) 'the itch of over-valuing foreign parts', although her journal was not published in full until 1888, with Robert Southey only publishing extracts from it in 1812.[20] Early in the eighteenth century, however, Daniel Defoe famously published his *Tour Through the Whole Island of Great Britain* in three volumes between 1724 and 1727. Despite such excursions, it was really only with the publication, as travel guides, of William Gilpin's *Observations* of various parts of Britain in the 1780s that tourism in England was to become a notable phenomenon.[21] Pennant's tours of Scotland in 1769 and 1772, therefore, were part of a continuum of travels focusing on the local and antiquarian, but the way he recounts his encounters with the *art* of the 'ancients' of Britain during his tours is notable, marking a clear break both with what had gone before and what was to emerge at the end of the century when such art works began to acquire a greater profile.

Camden's *Britannia* and its many spin-offs had more than provided for those seeking the antiquities of the 'Classical' (Roman) world locally. Defoe, for instance, had carried Gibson's 1722 revised edition with him as a reference book on his travels. In his account of Penrith in Cumbria, for example, he found 'several remarkable things', some of which were 'mentioned by the right reverend *continuator* of Mr. Cambden [emphasis added]', who – happily for Defoe – confirmed his own observations of:

> Two remarkable pillars fourteen or fifteen foot asunder, and twelve foot high the lowest of them, though they seem equal. The people told us, they were the monument of Sir Owen Caesar, the author above-nam'd calls him Sir Ewen Gesarius, and perhaps he may be right; but we have no inscription upon them. This Sir Owen, they tell us, was a champion of mighty strength, and of gygantick stature, and so he was, to be sure, if, as they say, he was as tall as one of the columns, and could touch both pillars with his hand at the same time.[22]

Anyone reading this can perhaps be forgiven for not immediately realizing that Defoe was referring to a collection of tenth-century Anglo-Scandinavian monuments, generally known as 'The Giant's Grave' that still stand in the churchyard in Penrith.[23] Defoe's account makes it clear why they have received this tag, but gives no indication that they might date to the tenth century. Rather, by implication, and on the authority of Camden, they are situated within the context of Romano-British legend. In fact, their Anglo-Saxon identity remained unrecognized until 1743 when Charles Lyttelton presented a paper on them to the Society of Antiquaries of London following his own travels through the island.[24]

The point is, that although tours through Britain, and particularly England became a recognized and more common phenomenon in the course of the eighteenth century, at the end of the century they came to be newly inspired by the search for what Gilpin had popularized as the 'Picturesque': that which was 'expressive of that peculiar kind of beauty, which is agreeable in a picture'.[25] When monuments of the vernacular ancient (early medieval) variety were encountered in the course of antiquarian-inspired travels, they were only rarely noted; for the most part they were either ignored (as 'curiosities of art'),[26] or were recorded for reasons that did not credit their origins. Against this background Pennant's treatment of the antiquities of early medieval Scotland is distinctive in the way it bridges different modes of perceiving the material remains of an ancient history and those of the landscape in which they could be encountered.[27]

Travel, Taste and the Antique: Thomas Pennant in Scotland

In fact it is only in the publication of Pennant's *Tours of Scotland* that we find *local* vernacular monuments of antiquity, the early medieval sculptures, receiving any sustained attention in their own right. And indeed, his accounts provide a wealth of information and insight to attitudes to such 'local' antiquities.[28] Following earlier descriptions of the region by Robert Sibbald, published in 1710 and 1711,[29] and that of Alexander Gordon, which had appeared in 1726,[30] both of which presented the Roman, 'Saxon' and Danish historical phases of the region in keeping with the chorographic tradition set out by Camden, Pennant's *Tours* of 1769 and 1772, which charted an actual tour that could be followed, were favourably received. The first (published in 1771), for instance, inspired Johnson to make his own tour of the Hebrides, and the second (published in 1774 and 1776) influenced his subsequent published account.[31] Furthermore, although Pennant's tours were undertaken from the point of view of a renowned natural scientist, he nevertheless recorded much that was neither botanical nor zoological. His journey through Cumberland, en route to Scotland, for instance, includes comments on the landscape. Leaving Thirlmere, he recounted how the countryside provided 'a strange and horrible view downwards, into a deep and misty vale, at this time appearing bottomless, and winding far amidst the mountains, darkened by their height, and the thick clouds that hung on their summits'.[32]

Alongside such atmospheric stage setting the early monuments of local antiquity encountered by Pennant provided him with objects relevant for what he termed 'accurate observation'. He presents Defoe's 'Giant's Tomb' at Penrith, for example, by means of taking the earlier illustration by Hugh Todd,[33] which Pennant and the 'gentlemen of the place whom [he] consulted', were convinced was 'entirely fictitious'. He then compared it for his readers with his own 'representation of those pillars [that] wanted the accuracy [he] wished', 'after a drawing by Moses Griffith' (Fig. 4.1).[34] He then proceeded by recording the measurements of the shafts and hogbacks, as well as the third shaft at Penrith, known locally as the 'Giant's Thumb' (which had not been noted by Camden, his 'continuator', or Defoe).[35] Pennant deemed these monuments to be 'evidently Christian, as appears by the cross on the cap'. Setting out his detailed observations, measurements and reasoned deductions he concluded by describing Camden's *Britannia* as a 'fable', because it considered the Penrith monuments to have been set up:

Figure 4.1 Pennant's illustration of *The Giant's Grave*, by Moses Griffith (below), compared with that published by Hugh Todd (above), from *A Tour in Scotland, MDCCLXIX* (Chester, 1771).

> to perpetuate the memory of *Cesarius*, a hero of gigantic stature, whose body
> extended from stone to stone: but it is probable that the space marked by these
> columns contained several bodies, or might have been a family sepulchre.[36]

The method of 'accurate observation' that Pennant brought to bear at Penrith
in the context of previous accounts, and his attempt to situate the monuments

in a perceived historical context removed from local fable, is the approach he continued to follow in respect of the ancient monuments he encountered at sites where the early and Christian activities were more historically une-quivocal than was the case at Penrith. On Iona, for instance, he recounts the Columban history of the site, naming Bede as his source, and presents the carved stone crosses and their remains according to their measurements, leaving them within a defined early Christian landscape.[37] Once he moved beyond such well-known *ecclesiastical* sites, however, Pennant discussed the early medieval carved monuments in a notably different manner, and one that does not vary throughout the *Tours*: he explained them exclusively as memorials to Scottish victories over invading Danes. In this he was likely influenced by Gordon who had also presented the early sculptures as memorials of military engagements with the Danes, but had done so in such a way that they functioned as 'markers' of the various events recounted in his narrative of the 'Danish period' of Scottish history, the presentation of which was the primary aim of the section.[38]

Thus, following Gordon and looking back to Hector Boyce's early sixteenth-century *Historia Gentis Scotorum*,[39] Pennant describes the cross at Camus, in Angus, as:

a curious monumental stone, set up in memory of the defeat of Camus, a Danish commander, slain on the spot, about the year 994 [which] is in the form of a cross.

Unlike Gordon, he then provides its measurements and, *after* describing its decoration, concludes with an account of the battle which 'was fought near the village of Barray; where numbers of Tumuli mark the place of slaughter. But Camus, flying, was slain here.'[40] Likewise, at Aberlemno, about 20 miles north of Camus, he records the cross slabs and notes that they are also 'supposed to have been erected in memory of victories over Danes; and other great events that happened in these parts' (Figs 4.2a and b). He then goes on to comment that 'These are *local* monuments, being unknown in Ireland, and indeed limited to the eastern side of North Britain [emphasis added]'.[41] Furthermore, he informs his readers that the cross slabs 'succeeded' the Pictish Symbol stones, which he regarded as being 'as artless as any of the old British monuments, which I apprehend from their excessive rudeness, [to be] the first efforts of the sculptor, and his success is such as might be expected'.[42] Compared with their 'artless' predecessors the cross slabs exhibited 'a fancy and elegance that does credit to the artists of those early days'.[43] With the chronology and aesthetics of the early Christian monuments thus established, he goes on to discredit the supposition (perpetuated in Sibbald's *Histories*),[44] that they have their origins

Figure 4.2a Road-side cross-slab, Aberlemo, Angus.

Figure 4.2b Pennant's illustration of the road-side cross-slab at Aberlemno, Angus (detail), from *A Tour in Scotland, MDCCLXXII, Part II* (London, 1776).

in Egypt, declaring that 'the historian's vanity in supposing his countrymen to have been derived from that ancient nation, is destitute of all authority';[45] he is likewise unconvinced by Bishop Nicolson's argument of 1702 that they should be compared 'to the [pagan] runic stones in Denmark and Sweden'.[46]

It is important to note here that Pennant's account of the early medieval monuments, which occupies over 40 successive pages, does *not* appear to follow the itinerary of his travels, and involves cross slabs set up across the length and breadth of Scotland. Thus, departing from his account of the monuments encountered in Angus, he 'accurately observes' the northern monuments at Forres (on the Moray coast, near Inverness); Dunfallandy (in Perthshire, central Scotland); and Glamis (back in Angus, in the south-east). Here, he pauses to note that this too is a memorial, this time 'erected in memory of the assassination of King Malcom' by the Danes – an event he recounts in some detail.[47] He then continues discussing the stones at Miegle (north and east of Glamis, in Perth and Kinross), before returning to Sueno's Stone at Forres on the Moray Firth. The name and decoration of this particular monument he suggests 'probably satirically allude[s] to the name of Sueno, or the Swine, a Danish monarch'.[48] He then links this to the Doctan Pillar (as he calls it), situated back in Angus, about 4 miles from Kirkaldie (at St Vigean's),[49] which he sent his 'servant, Moses Griffith' to draw, in order to correct the version 'erroneously figured by Sir Robert Sibbald' (Fig. 4.3). This he did, because it too was 'erected in memory of a victory, near the Leven, over the Danes in 874, under the leaders Hungar and Hubba, by the Scots, commanded by their prince, Constantine II'.[50]

By invoking the early medieval stones of Scotland in this distinctive manner, Pennant presents not so much an account of monuments encountered haphazardly in the course of his tour, but rather a consistent account of an ancient Scottish history: one characterized by independence and (implied Christian) victory in the face of foreign (pagan) invasion and occupation. In part, this is achieved by focusing on the carvings of hunting scenes, warriors and depictions of conflict and combat featured on the sculptures, rather than on the relief crosses preserved on the carved slabs, which have given the monument type its name in modern scholarship. While Gordon had recorded this type of Christian subject matter in his account,[51] Pennant notably glosses over this aspect of the decoration. Furthermore, rather than presenting the sculptures as markers of the various Scottish campaigns against the Danes, as Gordon does, Pennant focuses on the monuments in their own right with the historical setting contextualizing them. Published only 25 years after the Hanoverian defeat of Jacobitism in 1746, it is perhaps unsurprising that this viewpoint is presented implicitly and cumulatively through the apparently traditional vehicle of the antiquarian account of 'curiosities of art'. It certainly

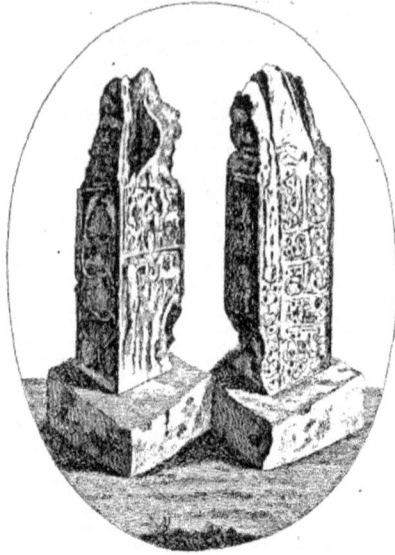

Figure 4.3 Pennant's illustration of the Drosten Stone, St Vigean's, Angus, from *A Tour in Scotland, MDCCLXXII, Part II* (London, 1776).

presents a divergence from the otherwise Hanoverian views expressed in the *Tours*, and perhaps reflects Pennant's accommodation of the earlier accounts with personal observation and a perceived need to promote the vernacular antiquities of early medieval Scotland.[52]

Whether this was indeed the case, the impact of this distinct, and at the time novel, presentation is hard to overestimate. Indeed, Charles Cordiner was to take Pennant's campaign even further, first in his *Antiquities and Scenery of the North of Scotland* (a series of letters addressed to Pennant), which was published in 1780, and then in his *Remarkable Ruins and Romantic Prospects*, published as two volumes between 1788 and 1795. Alongside accounts of the natural history of the region, and illustrations of the landscape and its ruins (painted by Cordiner himself, a student at the Academy of Fine Arts in Glasgow), his publications are filled with written accounts and illustrations of the early medieval carved stone monuments. For Cordiner, it is clear that:

these obelisks seem to have been shrines raised by the living, as a monument of their fame and power, and perhaps more frequently so, than as any memorial of the dead; for there never is any emblem of mortality on them.[53]

Their presumed function as 'monuments of victory and triumph' is rendered absolute by the manner in which many are illustrated: in the visual tradition of the 'trophy' – rather than as 'ruins in a landscape' (Fig. 4.4). This is a specific visual presentation, derived from the iconography of triumph in late antique imperial art, and clearly distinct from that more usually invoked, as for instance by Moses Griffith for Pennant (see, e.g. Fig. 4.2b), or indeed by J. M. W. Turner in his illustration of the crosses at Whalley in Lancashire, produced for Thomas Whitaker's *History of the Parish of Whalley* in 1800.[54]

Given the impact of Pennant's statements, verbal and visual, concerning the ancient statuary of early Christian Scotland, comments on the

Figure 4.4 Cordiner's illustration of *Symbols of Caledonian Monuments compared with some on the Pamphilian Obelisk at Rome*, from Cordiner, *Remarkable Ruins* (London, 1795).

'ancient' monuments elsewhere in Britain might be expected to be more forthcoming at this time, after their relative obscurity in previous accounts of vernacular antiquities. This however, is not the case. It is notable in England, for instance, that while perceptions of the national characteristics and political mores of the Anglo-Saxons were increasingly positive with the rise of Whig perceptions of their history, little is said about their Christian affiliations in the texts promoting the ancient German forefathers of the English laws and nation; it was only their pagan beliefs that were explicated. It is an emphasis that was established, clearly, in Milton's 1670 *History of Britain*.

While there are a number of possible explanations for this 'absence', it may reflect contemporary attitudes towards Rome. Although the pronouncements of men like Winckelmann and Johnson concerning the art and culture of the Romans were entirely favourable, politically, 'the Romans' were not always viewed positively,[55] and the Church of Rome was of course still liable to a bad press, with treatises, such as those by Gibson, being published on how to avoid 'the growth and mischiefs of popery'.[56] In part, therefore, the carved stone sculptures of Anglo-Saxon England, understood to be Christian by their form and decoration, may still have been perceived to be Roman Catholic (albeit 'English'); they were thus not monuments that could be easily discussed in terms other than their inscriptions. These were the features of Anglo-Saxon sculpture that had inspired the earliest interest in the material in the sixteenth century, and continued to be the focus of discussion through into the twentieth.

In this respect William Cole's account of a fragment of the larger of the Sandbach crosses in 1755 is quite remarkable. Cole had flirted with Catholicism when in France at the start of his Grand Tour, but returned to England before going any further (ecclesiastically or geographically), and took up the living at Milton in the fens just north of Cambridge where he spent the rest of his life. It was when staying in Tarpoley in Cheshire with the Revd John Allen, 'a great Admirer of venerable Antiquity', that he encountered the 'most curious and antient Cross opposite the Kitchin Door by the Pump, in an obscure Place by the necessary House'.[57] His account of this curiosity is preceded by a lengthy description of its erstwhile owner, Sir John Crewe, who had died in May 1755. He is commemorated by Cole as:

> a very sensible Man & a good Antiquarian, descended [from] the First Lord Crewe [who] was a Lover of the Constitution both in Church and State, and consequently an Enemy of Popery and arbitrary Government: [and was] stedfast to the establisht Religion.[58]

Cole goes on to note that John Crewe himself, in acquiring this piece of the Sandbach cross, 'had very hard Mortar plaistered over' the panel of 'our B.Saviour on the Cross, which, as a Relique of Popery, Sir John Crewe look'd upon as improper to be publicly seen, lest it should be worshipped'.[59] This plaster having been 'with Difficulty, taken off again', Cole was able to describe the crucifixion scene, but (perhaps understandably) experienced problems of interpretation: the angel over the Nativity thus becomes an Eagle, while the figures above the crucifixion, comprising the Adoration of the Magi, are identified as 'God the Father who is sitting, and before him our B.Lady with her Babe'. The account is written out over two recto folios of his notebook, but on the intervening verso Cole has transcribed William Smith's late sixteenth-century account of the crosses in the *Vale Royal*, which records 'writings', indecipherable unless someone 'be holden with his head downwards'.[60]

What Cole preserves in his account is both the continued interest in the epigraphy (and local traditions) related to the Anglo-Saxon monuments, and also the manner in which they were still associated with 'popery'. His declarations of being unable to identify much of what he sees, proclaiming that 'it wants the sagacity & Ingenuity of a Dr. Stukeley to ascertain the Reality of each Figure', are particularly interesting in this respect: his drawing of the carvings that faces the opening of the written account presents a very clear representation of the figures above the Crucifixion, as the Virgin and Christ Child faced by three bust-length figures (Fig. 4.5). This suggests that in the middle of the eighteenth century, while the traditions, stone type, measurements and epigraphy of local (Anglo-Saxon) monuments of antiquity could be recorded with impunity, and the non-figural decoration of early Medieval Christian sculptures, the 'Knots, Foliages & Birds & Quadrupeds', could be declared 'very pretty',[61] the overtly Christian subject matter of the figural reliefs was a fraught object of enquiry. Unlike the monuments of vernacular antiquity encountered by Pennant in Scotland, those surviving from Anglo-Saxon England that were accessible in the eighteenth century tend not to depict scenes of warriors and combat, and generally lack local traditions linking them to local victories and defeats.[62] It was exactly these aspects of the Scottish sculptures that had enabled Pennant to ignore or downplay the high-relief crosses and their associated Christian figural schemes on the monuments he was discussing, and re-present them as memorials commemorating local military encounters and victories over foreign invaders. Here it is perhaps possible to see him responding in a manner analogous to that presented by Cole, and in keeping with the disquiet that the Catholicism of medieval monuments inspired in him elsewhere – notably at St Winefride's Well in Flintshire, which Mary-Ann Constantine has recently elucidated.[63]

The learned & worthy Br. Gibson in his Edition of Camden's Britannia in Flintshire p. 829, 830, seems to be at a Loss to account for a remarkable Monument or carved Pillar on Thos tyn Mountain, which he has given us a double Print of in the Plate at the End of the Account of Wales. This Pillar, I take, to be no more than a Cross set up for Devotion & not unlikely near the Place of Sepulture after some Battle. That it is not Runic or Danish, I think, is evident by the Crofes on both Sides at Top; as also by another in the midst of one of the Fronts. I make no Doubt but it was a Cross set up on some great Solemnity & in Commemoration of some remarkable Victory or Battle near the Place: & the Sides of it are curiously carved & engraved, in the same manner as those at Torporley & Sandbach, by having Eblinges or Knots carved on them. It is not at all improbable but those Pillars at Torporley & Sandbach were terminated at Top by a rounded Head in which might be carved a Cross flory, as in that of Mostyn Mountain: for I have observed that the Top of this Cross is broken off.

Figure 4.5 William Cole, drawing of the upper stone of the North Cross, Sandbach Market Place, Cheshire (1775).

Whether this is indeed the case, it is at exactly this point, when political opinions were shifting, when Taste was being cultivated, and when perceptions of the arts of antiquity and the ancient past were being placed centre stage, that points of view seem to have emerged which allowed scholars to free the Anglo-Saxon sculptural monuments from their Christian/papist background: perceptions that effectively neutralized them and at the same time had the effect of elevating them (by association), to the status of the art of Classical antiquity. Against this background, Pennant provides a clear insight as to how such processes of transition could occur; namely, through the presentation of his use of the term 'obelisk'. As I have argued elsewhere, this was a term used throughout the Middle Ages and into the seventeenth century to denote monuments of any form that were understood to function as funerary monuments; the term 'obelisk' never really denoted (when referring to the early medieval monuments of Britain) the *form* of the sculpture.[64] But as we have seen, from the mid-eighteenth century 'obelisks', especially those in Rome, were accruing new meanings: meanings that involved triumphal monumentalism. This was partly due to the widespread publicity surrounding the excavation and re-erection of the obelisk of Augustus' Horologium which, coincidentally, was re-erected with sections of granite taken from the triumphal column of Antoninus Pius.

In the light of this, it is perhaps no coincidence that, in addition to his declaration that the early medieval monuments of Scotland were obelisks but not funerary monuments,[65] Cordiner saw fit to compare one of his illustrated trophies of Scottish 'obelisks', erected in memory of victories over the Danes, with the 'Pamphilian Obelisk' in Rome.[66] Earlier, however, in his second *Tour of Scotland* in 1772, Pennant had published an account of the cross he encountered at Ruthwell in Dumfriesshire, which he signalled with a marginal note: 'Ancient Obelisk'.[67] This was done in such a way that the annotation, and therefore the monument itself, stands out on the page. Having identified one of the cross arms of the monument at Ruthwell, Pennant is in no doubt that the obelisk is a cross, and indeed refers to it as such at the end of his account, following his usual careful provision of the measurements, the various fragments within the church, their Latin inscriptions,[68] and what he terms the 'rude' figural carvings – easily identifiable as depicting biblical and Christian subject matter which, unlike Gordon, he does not recount in detail.[69] While Gordon, in 1726, following the long-standing tradition of seeing such monuments as obelisks due to their perceived funerary memorial function, had referred to the Ruthwell monument as an obelisk 'in Form, like the Aegyptian Obelisks at Rome',[70] this comment is textually contained rather than highlighted in the layout of the page, and nowhere does he mention that it was a cross. By contrast, the overall effect of the presentation of Pennant's account is to downplay the Christian nature of the monument, despite his awareness of its Christian form, and emphasize its 'Classical' nature. This is in direct contradistinction to the terminology he uses for the other early medieval monuments presented in the course of the *Tours*, which are variously denoted: as grave-stones (at Glamis, Miegle and Aberlemno), a curious monumental stone (at Camus), columns (at Penrith), and crosses (as at Iona) – terms which seem, in context, to indicate that Pennant was deliberately avoiding the use of the word obelisk, preserving it for the monument at Ruthwell to accentuate his perception of its 'Classical' antiquity, perhaps inspired by its use of lengthy Latin inscriptions.

Conclusion

During the course of the eighteenth century, in the context of the tours undertaken across the Continent and particularly to Italy, there seems to have been a shift in perception of what an obelisk could be, from funerary markers to triumphal monuments in Classical Roman antiquity. Pennant provides a nuanced insight into the way in which the term could be applied to the early medieval monuments in Britain, thus potentially freeing them for discussion as examples of monumental (triumphal) art, rather than monuments of superstition. Following Pennant's publications, the early medieval monuments of Britain were discussed more frequently, and in ways that implicitly invoked the principle set forward by Winckelmann, that 'reasonable imitation takes just the hint'.[71] This is certainly the attitude with which Pennant re-presents the vernacular monuments of 'the ancients' in Scotland. In his descriptions he simultaneously invokes them to demonstrate Taste and advance an objective and scientific methodology of explanation. Furthermore, he enables them to be identified as something in their own right: memorials of ancient Scotland.

Notes

1 E.g., Thomas Nugent, *The Grand Tour containing an Exact Description of most of the Cities, Towns and Remarkable Places of Europe by Mr [Thomas] Nugent*, 4 vols (London, 1743), passim.
2 Johann Joachim Winckelmann, *Gedanken über die Nachahmung der griechischen Werke in Malerei und Bildhauerkunst* (Dresden und Leipzig, 1755; 2nd ed. 1756).
3 Henry Fuseli, *Reflections on the painting and sculpture of the Greeks: with instructions for the connoisseur, and an essay on grace in works of art. Translated from the German original of Abbé Winkelmann, Librarian of the Vatican, F. R. S. &c. &c*, 2nd ed. (London, 1767), 30.
4 Ibid. 2.
5 Ibid. 256.
6 Ibid.
7 Ibid. 257.

8 See, e.g., Rudolf Wittkower, 'Imitation, Eclecticism, and Genius', in *Aspects of the Eighteenth Century*, ed. Earl R. Wasserman (Baltimore, MD: John Hopkins Press, 1965), 143–61; James L. Larson, 'Winckelmann's Essay on Imitation', *Eighteenth-Century Studies* 9, no. 3 (1976): 390–405. Winckelmann's *Geschichte der Kunst des Alterthums* (Dresden, 1764), was regarded as providing 'a thorough, comprehensive and lucid chronological account of all antique art' (Francis Haskell and Nicholas Penny, *Taste and the Antique: The Lure of Classical Sculpture, 1500–1900* [London and New Haven, CT: Yale University Press, 1981], 101), although not translated into English till 1849, by G. Henry Lodge; see also Joan DeJean, *Ancients against Moderns: Culture Wars and the Making of a Fin de Siecle* (Chicago, IL: Chicago University Press, 1997).

9 See, e.g., Walter Pater, *Winckelmann* (Heidelberg: Universitätsbibliothek der Universität Heidelberg, 1867); Denis M. Sweet, 'The Personal, the Political, and the Aesthetic: Johann Joachim Winckelmann's German Enlightenment Life', *Journal of Homosexuality* 16, nos. 1–2 (1988): 147–62; Édouard Pommier, *Winckelmann: la naissance de l'histoire de l'art à l'époque des Lumières: actes du cycle de conférences prononcées à l'Auditorium du Louvre du 11 décembre 1989 au 12 février 1990* (Paris: La Documentation Française, 1991); Alex Potts, *Flesh and the Ideal: Winckelmann and the Origins of Art History* (New Haven, CT: Yale University Press 1994).

10 Rosemary Sweet, *Cities and the Grand Tour: The British in Italy, c.1690–1820* (Cambridge: Cambridge University Press, 2012).

11 James Boswell, *The Life of Samuel Johnson, LL.D*, 2 vols (London, 1791), II: 61. For Johnson's abortive Grand Tour plans, see e.g. Peter Martin, *Samuel Johnson: A Biography* (London: Weidenfeld and Nicolson, 2008), 457–65.

12 James Stuart, *De Obelisco Caesaris Augusti e Campo Martio nuperrime effosso* (Rome, 1750).

13 Biblioteca Medicea Laurenziana, Florence, Amiatinus 1; see Jonathan J. G. Alexander, *Insular Manuscripts: 6th to the 9th Century* (London: Harvey Miller, 1978), 32–35, cat. 7. Bandani moved to the Laurenziana in Florence in 1756; see Frank Salmon, 'Stuart as Antiquary and Archaeologist in Italy and Greece', in *James "Athenian" Stuart. 1713–1788: The Rediscovery of Antiquity*, ed. Susan Weber Soros (New Haven, CT and London: Yale University Press, 2006), 103–45 (114).

14 Catherine Arbuthnott, 'The Life of James "Athenian" Stuart, 1713–1788', in *James "Athenian" Stuart*, ed. Soros, 59–101 (66).

15 Salmon, 'Stuart as Antiquary', 103.

16 See, e.g., Colin Thomas, 'Thomas Pennant 1726–1798', in *Geographers: Biobibliographical Studies* 20, ed. Patrick H. Armstrong and Geoffrey J. Martin (London: Bloomsbury Publishing, 2000), 85–101 (87); Jeremy Black, *The British and the Grand Tour* (London: Croom Helm Ltd., 1985), 167.

17 Bernard Nurse, 'The 1610 Edition of Camden's *Britannia*', *Antiquaries Journal* 73 (1993), 158–60 (159); see also Robert Mayhew, 'Edmund Gibson's Editions of *Britannia*: Dynastic Chorography and the Particularist Politics of Precedent, 1695–1722', *Historical Research* 73, no. 182 (2000), 239–61 (239); for summary, see Jane Hawkes, 'Creating a View: Anglo-Saxon Sculpture in the Sixteenth Century', in *Making Histories*, ed. eadem (Donington: Shaun Tyas, 2013), 372–84.

18 See, e.g., Stuart Piggott, *William Stukeley: An Eighteenth-Century Antiquary* (Oxford: Clarendon Press, 1950), 207; idem, *Ruins in a Landscape: Essays in Antiquarianism* (Edinburgh: Edinburgh University Press, 1976), 12; and more recently the essays in Martin Myrone and Lucy Peltz, eds, *Producing the Past: Aspects of Antiquarian Culture and Practice, 1700–1850* (Aldershot: Ashgate Publishing, 1999); for summary see Hawkes, 'Creating a View'.

19 Rosemary Sweet, *Antiquaries: The Discovery of the Past in Eighteenth-Century Britain* (London: Hambledon & London, 2004).

20 Celia Fiennes, *Through England on a Side Saddle in the Time of William and Mary, Being the Diary of Celia Fiennes*, with introduction by Emily Wingfield Griffiths (London: Field & Tuer, 1888), n.p.; Robert Southey, ed., *Omniana: or, Horae Otiosiores* (London: Longman, 1812); see Nat Lewis Kaderly, 'Southey's Borrowings from Celia Fiennes', *Modern Language Notes* 69, no. 4 (April 1954): 249–53; see also Richard Lassels, *The Voyage of Italy, or A compleat journey through Italy. In two parts* (Paris, 1670).

21 Gilpin's *Observations* were variously published in 1782 (on the River Wye and South Wales), 1789 (on several parts of Britain, but particularly the Scottish Highlands), 1792 (on Cumberland and Westmoreland), a publication supplemented in 1798 (on Western England and the Isle of Wight); his *Remarks, on the New Forest in Hampshire* had been published in 1791. See also Karl Philipp Moritz, *Travels, chiefly on Foot, through several parts of England in 1782, described in Letters to a Friend* (1758; London, 1795), a clergyman/student who travelled through England with his copy of *Paradise Lost* in order to read it 'in the land of Milton'.

22 Daniel Defoe, *A Tour Through the Whole Island of Great Britain*, 7th ed., 4 vols (London, 1769), III: 306.

23 Richard N. Bailey and Rosemary J. Cramp, *Cumberland, Westmorland, and Lancashire-North-of-the-Sands* (Oxford: Oxford University Press, 1988), 136–39.

24 Dr [Charles] Lyttelton, 'An Account of a Remarkable Monument in Penrith Church-yard, Cumberland', *Archaeologia* 2 (1773): 48–53; for travels, see SAL, 187h, f. 63.

25 William Gilpin, *An essay upon prints: containing remarks upon the principles of picturesque beauty* (London, 1768), x; see also idem, *Observations on the River Wye, and several parts of south Wales &c.* (London, 1784), 1–2:

> We travel for various purposes; to explore the culture of soils; to view curiosities of art; to survey the beauties of nature; to search for her productions; and to learn the manners of men; their different polities, and modes of life. The following [...] proposes a new object of pursuit; that of not barely examining the face of a country; but of examining it by the rules of picturesque beauty; that of not merely describing; but of adapting the description of natural scenery to the principles of artificial landscape; and of opening the sources of those pleasures, which are derived from the comparison.

26 Gilpin, *Observations on the River Wye*, 1.

27 See discussion of Pennant's analogous treatment of history and the landscape in Wales in Mary-Ann Constantine, '"To trace thy country's glories to their source": Dangerous History in Thomas Pennant's *Tour in Wales*', in *Rethinking British Romantic History, 1770–1845*, ed. Porscha Fermanis and John Regan (Oxford: Oxford University Press, 2014), 121–43 (126).

28 Thomas Pennant, *A Tour in Scotland; MDCCLXIX* (Chester, 1771); idem, *A Tour in Scotland, and Voyage to the Hebrides, MDCCLXXII* (Chester, 1774); idem, *A Tour in Scotland, and Voyage to the Hebrides, MDCCLXXII, Part 2* (London, 1776).

29 Robert Sibbald, *The History, Ancient and Modern, of the Sheriffdomes of Fife and Kinross; with The Description of Both and of the Firths of Forth and Tay, and the Islands in Them* (Edinburgh, 1710); idem, *Portus; Coloniae & Castella Romana [...] or, Conjectures Concerning the Roman Ports, Colonies, and Forts, in the Firths* (Edinburgh, 1711).

30 Alexander Gordon, *Itinerarium septentrionale: or, a journey through most of the counties of Scotland and those in the north of England* (London, 1726).

31 *Critical Review* (January 1772): 28, described Pennant's *Tour* (1771) as 'the best itinerary which has hitherto been written on that country'; see James Boswell, *The Journal of a Tour to the Hebrides, with Samuel Johnson* (Dublin, 1785); idem, *Life of Samuel Johnson*, II: 215–16.

32 Pennant, *Tour in Scotland […] 1772*, I: 42; see also account of Derwentwater, ibid. 45–46.

33 See D. J. W. Mawson, 'Dr Hugh Todd's Account of the Diocese of Carlisle', *Transactions of the Cumberland and Westmorland Antiquarian and Archaeological Society*, new series, 88 (1988): 207–24; idem, ODNB *s.n.* Todd, Hugh (*c.*1657–1728) (accessed 25 August 2014).

34 Pennant, *Tour in Scotland […] 1772*, I: 274–75, Plate XIII.

35 Bailey and Cramp, *Cumberland*, 135–36, 484–88.

36 Pennant, *Tour in Scotland […] 1772*, I: 274.

37 Ibid. 291.

38 See, e.g., Gordon's account of the stones at Aberlemno, Camus and Forres, *Itinerarium septentrionale*, 151, 154, 158.

39 Hector Boyce, *Scotorum historiae a prima gentis origine*, (Paris, 1575).

40 Pennant, *Tour in Scotland […] 1772*, II: 129; see Gordon, *Itinerarium septentrionale*, 154, Plate 54.

41 Pennant, *Tour in Scotland […] 1772*, II: 167.

42 Ibid. 168.

43 Ibid.; compare account with that of Gordon, *Itinerarium septentrionale*, 151–52, Plate 53.

44 Sibbald, *Sheriffdomes of Fife and Kinross*.

45 Likely referring to account by Boyce, *Scotorum historiae*; see similar passage in Gordon, *Itinerarium septentrionale*, 159.

46 Pennant, *Tour in Scotland […] 1772*, II: 169; William Nicolson, *The Scottish historical library: containing a short view and character of most of the writers, records, registers, law-books, &c. which may be Serviceable to the Undertakers of a General History of Scotland* (London, 1702), 65. Pennant's opinion notably differs from that of his predecessors, such as Alexander Gordon, *Itinerarium septentrionale*, 159, who, referencing 'Bishop Nicholson in his Scots Historical Library page 64', agrees with the Scandinavian comparison.

47 Pennant, *Tour in Scotland […] 1772*, II: 174; compare with Gordon, *Itinerarium septentrionale*, 162, Plates 61.2–3.

48 Pennant, *Tour in Scotland […] 1772*, II, 174; compare with Gordon, *Itinerarium septentrionale*, 58–59, Plate 56.

49 Now referred to as the Drosten Stone.

50 Pennant, *Tour in Scotland […] 1772*, II: 205.

51 See, e.g., Gordon's account of the Crucifixion and Evangelists featured at Aberlemno, *Itinerarium septentrionale*, 151.

52 See discussion in Constantine, '"To trace thy country's glories to their source"', 124–30. I am grateful to her for advice on this subject.

53 Charles Cordiner, *Remarkable ruins, and romantic prospects, of North Britain, with ancient monuments, and singular subjects of natural history*, 2 vols (London, 1788–95), I: n.p.

54 Thomas Dunham Whitaker, *An history of the original parish of Whalley, and honor of Clitheroe, in the counties of Lancaster and York* (Blackburn: Printed by Hemingway & Crook, 1801).

55 See, e.g., George B. Clarke, 'Grecian Taste and Gothic Virtue: Lord Cobham's Gardening Programme and its Iconography', in *Apollo: Patrons and Patriots, I. The Rise of a National School*, ed. D. Sutton (September 1985), 566–71 (570).

56 Edmund Gibson, *A sermon of the growth and mischiefs of popery: preach'd at the assizes held at Kingston in Surrey, Sept. 5th, 1706* (London, 1706).

57 BL 5830, f. 34r.

58 BL 5830, ff. 16r–18r.

59 BL 5830, f. 34r.

60 BL 5830, f. 34v.

61 BL 5830, f. 35r.

62 For one notable exception to this, see the tenth-century shaft fragments at Checkley, Staffordshire, the figural decoration of which cannot be clearly associated with biblical or Christian subject matter; from the time of their recording, by Plot in 1686, they were associated as memorials to bishops slain in battle by the Danes; see Robert Plot, *Natural History of Staffordshire* (Oxford, 1686), 432.

63 Constantine, '"To trace thy country's glories to their source"', 130–33.

64 Hawkes, 'Creating a View', 372–84.

65 See above, 48n.

66 Cordiner, *Remarkable ruins*, II: unnumbered plate illustrating 'Symbols of Caledonian Monuments, compared with some on the Pamphilian Obelisk at Rome'.

67 Pennant, *Tour in Scotland [...] 1772*, I: 96.

68 Ibid. I: 96–97, assumes that the runic inscriptions on the narrow sides of the shaft indicate a second campaign of decoration added by the Danes, the initial monument with its Latin inscriptions being produced by the Saxons.

69 Ibid.

70 Gordon, *Itinerarium septentrionale*, 161.

71 See above, 5n (Fuseli, *Reflections*, 256).

Chapter 5

SHAPING A HEROIC LIFE: THOMAS PENNANT ON OWEN GLYNDWR

Dafydd Johnston

Owen Glyndwr[1] is recognized today as one of the great figures of Welsh history, the leader of a revolt which commanded widespread support and presented a serious threat to English authority in the early years of the fifteenth century, and a hero who has come to embody Welsh aspirations to nationhood. Although he has been the subject of substantial academic study, much about him remains enigmatic, and he is a source of continual fascination.[2] However, his status has not always been so high, and for over three centuries after his death he was the subject of legend but also a cause of shame for some of his countrymen. Thomas Pennant's account of his life in the first volume of his *Tour in Wales* was a major contribution to the historiography of Glyndwr and led to his rehabilitation as a national hero in the nineteenth century. This article will consider his use of a wide range of historical sources to shape his life of Glyndwr within the context of travel writing.

Born about 1359, Glyndwr spent much of his life as a prosperous landowner in north-east Wales, having trained in the law and done military service in Scotland in the 1380s. His revolt began with an attack on the town of Ruthin in September 1400, and spread throughout Wales in the following years, reaching its height in the years 1404 to 1406 when Glyndwr had support from France and was able to hold parliaments at Machynlleth and Harlech. From 1407 onwards the king's forces gradually reestablished control, but resistance was not finally suppressed until 1415, which is the probable year of Glyndwr's death.

Glyndwr had received relatively little attention in published histories of Wales before Pennant's time, partly because of the tendency to regard the later Middle Ages as a mere interlude between the age of the independent Princes and the Tudor Acts of Union. The brief account of his rebellion in David Powel's *Historie of Cambria* (1584), closely followed by William Wynne in his *History of Wales* (1697), is in a neutral chronicle style, in places openly critical of its effects. Patriotic history writing gained ground during the eighteenth century, and a more celebratory attitude to Glyndwr is evident in Evan Evans's poem *The Love of Our Country* (1772) which depicts the brave resistance of 'great Glyndwr' to the usurper Henry.[3] The first biographical account of the man himself was 'Memoirs of Owen Glendower' published by John Thomas in 1775 as an appendix to his *History of the Island of Anglesey*, but probably in fact the work of the seventeenth-century antiquarian Robert Vaughan (see below). This may well have stimulated Pennant to publish a fuller and more sympathetic account of Glyndwr's life three years later.

Pennant's account of Glyndwr begins with an excursion westwards from Llangollen, crossing a significant county boundary, as the *mise en page* makes clear:

> After a descent of no great length, enter
> MERIONETHSHIRE,
>> into that portion for ever to be distinguished in the Welsh annals, on account of the hero it produced, who made such a figure in the beginning of the fifteenth century. This tract was antiently a comot in the kingdom of Mathraval, or Powys; and still retains its former title *Glyn-dwrdwy*, or the valley of the Dee. It extends about seven miles in length; is narrow, fertile in grass, bounded by lofty hills, often cloathed with trees; and lies in the parishes of Llangollen, Llandysilio, Llansantffraid, and Corwen.[4]

The county of Merioneth lay mainly in the north-west of the country, and consisted predominantly of remote and mountainous countryside which Pennant explored on his journey to Snowdonia recounted in the second volume of his *Tours*. On this occasion he ventured only as far as the site of Glyndwr's former home at Corwen at the most easterly tip of Merioneth before retracing his steps to Llangollen, as he informs his readers 69 pages later:

> Having now collected every thing in my power relating to this celebrated Briton, I return, by the same road, cross the Dee at Llangollen.[5]

Concise biographies of local worthies are a feature of Pennant's topographical writing which merit further attention as a method of linking history and place. Often inspired by paintings in the houses he visited, they usually occupy anything from a single paragraph to a few pages.[6] A typical example from his second Scottish tour is the account of the diarist and patron Lady Anne Clifford (1590–1676) which fills just over three pages.[7] The treatment of Owen Glyndwr, however, completely disrupts the geographical organization of the

Welsh tour, which is presumably why it was relegated to an appendix in the edition of 1810 edited by Pennant's son David.[8]

Owen's cognomen provided a compelling reason for placing the account of his life at this point in the tour, Glyndwr being an abbreviated form of the Welsh name for this part of the valley of the river Dee, Glyndyfrdwy (or 'Glyn-dwrdwy' as Pennant has it). And it was in any case a fundamental premise of travel writing that knowledge of history was best attained by personal observation and inquiry in its original location, which is why Pennant chose to imply that the material for his account had been collected in the course of his visit to Glyndwr's native region. But in fact he drew on a remarkably wide range of informants and sources to provide a history of national, and indeed international scope.

One of Pennant's great strengths as a travel writer was his ability to draw on the assistance of a network of like-minded antiquarians, from amongst his fellow gentry, such as Philip Yorke of Erddig and Paul Panton of Plas Gwyn in Anglesey, and from the Welsh clergy.[9] Although he acknowledges his debt to some of these in general terms in the 'Advertisement' to the first volume of his *Tour*, he does tend to present things in the body of the text as the fruit of personal observation which were in fact based on information provided by others.

His principal assistant and travelling companion on his Welsh tours was John Lloyd (1733–93), rector of Caerwys, who helped him in particular with Welsh-language material (Pennant himself having very little Welsh), transcribing and translating extracts from manuscripts and communicating with monoglot Welsh speakers.[10] Lloyd provided Pennant with a dossier on Glyndwr based on his research in private libraries such as Mostyn, Wynnstay, Hengwrt, Panton and even Pennant's own library at Downing.[11] Letters from Lloyd to Pennant in 1777 contain comments on drafts and show that Lloyd undertook fieldwork which Pennant drew on in his account:

> I made a shift to pay a visit to Corwen & its vicinity last week, but met with nothing material relative to Owen Glyndwr […] I visited Owen Glyndwr's prison, part of it is intire & consists of very strong heart of oak wooden Building & carries the signs of being coeval with Owen; from what is remaining the whole may be restored by the pencil of Moses [Griffith] and I think a drawing of it would not be improper to follow his life. I think besides that there is a paragraph or two that ought to be alter'd from a cursory view I made of 'Vita Ricardi 2d' at Mostyn. Next week I intend to dedicate to your Library at Downing.[12]

John Lloyd was a student at Oxford in the 1750s, and was able to draw on contacts that he had made there for his antiquarian research. One of those was Lewis Owen, rector of Llangelynin in Caernarfonshire, who was the source of information relating to Glyndwr's genealogy and birth from a manuscript in his possession.[13] Another antiquarian friend of Lloyd's from his Oxford days was John Thomas (1736–69) of Beaumaris, author of the *History of the Island of Anglesey*, published posthumously and anonymously in 1775, to which was appended 'Memoirs of Owen Glendowr', a seventeenth-century text attributed to Thomas Ellis, rector of Dolgellau and fellow of Jesus College Oxford, but likely in fact to have been the work of the antiquarian Robert Vaughan of Hengwrt.[14] Even though Pennant often follows the wording of the 'Memoirs' quite closely he makes no mention of it as a source, probably because he was using one or more copies of manuscript sources of that text prepared for him by John Lloyd.[15]

As the extract from one of his letters quoted above shows, Lloyd also gathered material on Glyndwr from printed sources, which can be traced in the footnotes to Pennant's *Tour*. The obvious starting point would have been the chronicles of Edward Hall (1548) and Raphael Holinshed (1587) which were among the main sources for Shakespeare's history plays. An earlier chronicle which was also published in the Tudor period was Thomas Walsingham's *St Alban's Chronicle*, written in 1423 and published in 1574. And a chronicle contemporary with Glyndwr himself was the *Historia Vitae et Regni Ricardi II* which covered the beginning of the rebellion up to 1402, and was available in an edition published by Thomas Hearne in 1729 (which Lloyd consulted in the Mostyn library; see extract from letter above).[16] Pennant's account also draws on Scottish and French chronicles from the fifteenth century, the *Scotichronicon* of Walter Bower,[17] and Enguerrand de Monstrelet's *Chronique*,[18] which offer an alternative perspective to that of the English chroniclers and serve to emphasize the international dimension of the rebellion.

In addition to these printed editions of chronicles, Pennant also had access to contemporary documents in Thomas Rymer's *Foedera*, a collection of official papers published in sixteen volumes between 1704 and 1713, which John Lloyd seems to have searched diligently for material relating to Glyndwr. And for decisions made by the Westminster Parliament he was able to consult what is referred to as 'Drake's *Parliamentary History*'.[19] The information from these two sources relates mainly to the English campaigns against Glyndwr, documenting the movements of the king and his son Prince Henry, and royal pardons granted to rebels. However, Pennant also quotes from Rymer's edition of letters written by Glyndwr authorizing his ambassadors to France in 1404.[20] And he gives details of an inquisition of rebels taken at Beaumaris in 1406 from a transcript of the original manuscript (both now lost) made by Edward Lhuyd (*c*.1660–1709), whose Welsh manuscripts he had been lent by Sir John Sebright.[21]

John Lloyd also advised on Welsh-language poetry as a source of information about Glyndwr. Five poems were used, all previously published. Three were in Rhys Jones's anthology, *Gorchestion Beirdd Cymru* (1773), where they are attributed to Iolo Goch.[22] One of these is supposed to describe the comet of 1402, and another to be a call for Glyndwr's return when he went into hiding in 1405.[23] The third, and the only one now considered to actually relate to Glyndwr, is Iolo Goch's description of his house and estate, of which Pennant gives a detailed summary presumably provided by Lloyd.[24] This is a very valuable addition to the account of Glyndwr's life, countering the hostile image of the English chroniclers by presenting him before his rebellion in a domestic setting with his wife and children and showing him to be a cultured gentleman respected for his generosity. Pennant relates the description to remains on the site of Glyndwr's residence in Corwen, where there happened to be a mound surrounded by a ditch and parkland as described in the poem. However, this is completely mis-leading, since the subject of Iolo Goch's poem was in fact another house belong-ing to Glyndwr at Sycharth in the parish of Llansilin near Oswestry, on the other side of the Berwyn mountains from his patrimony in Glyndyfrdwy. The error seems surprising to the modern reader, since Sycharth is now a very well-known historical site with clearly visible remains, but apparently it was not known in Pennant's time.[25] It is an understandable error given the similarity of the two sites and the importance Pennant attached to unity of place in Glyndyfrdwy.

The two other pieces of Welsh poetry used by Pennant had been pub-lished in appendices to John Thomas's text of the 'Memoirs'. One is a couplet from an elegy by Lewys Glyn Cothi for one of Glyndwr's illegitimate daugh-ters, Gwenllian, which Pennant quotes in Welsh with an English translation.[26] The other is a complete poem by Gruffudd Llwyd praising Glyndwr's martial prowess, which was published by Thomas with an English prose translation.[27] Pennant quotes only the opening line in Welsh and then gives a very free English version 'agreeably paraphrased by a bard of 1773'.[28] This was the work of the Revd Richard Williams (1747–1811) of Fron near Mold, another Oxford-educated clergyman who contributed literary translations of several Welsh poems to the *Tours*.[29] The original poem actually celebrated Glyndwr's exploits on behalf of the English crown in Scotland in the 1380s; the English version presents it as a prophecy of victory over King Henry made at the very beginning of the rebellion, but with one eye also on Glyndwr's posthumous reputation, as seen here in the concluding verse:

Strike then your harps, ye *Cambrian* bards!
The song of triumph best rewards
An hero's toils. Let *Henry* weep
His warriors wrapt in everlasting sleep;

> Success and victory are thine,
> *Owain Glyndwrdwy* divine!
> Dominion, honor, pleasure, praise,
> Attend upon thy vigorous days!
> And, when thy evening sun is set,
> May grateful *Cambria* ne'er forget
> Thy noontide blaze; but on thy tomb
> Never-fading laurels bloom![30]

The role of the bards as inspirers of the struggle for national liberty is a theme which frames Pennant's account of the Glyndwr rebellion, introduced here by Gruffudd Llwyd's poem heralding the beginning of the insurrection, and restated at its end in a patriotic peroration to be discussed below.

Glyndwr was able to claim widespread support in Wales on the basis of his royal lineage, being a direct descendant on his father's side from the princes of Powys and on his mother's side from the princes of Deheubarth in south-west Wales. He also had a more distant connection with the Gwynedd royal dynasty, and thus combined in his person the three main royal houses of Wales.[31] The contemporary view of his tripartite royal lineage is expressed in a poem by Iolo Goch,[32] but this poem does not seem to have been known to Pennant (or rather to John Lloyd), whose interpretation focuses solely on descent from the princes of Gwynedd. Pennant follows the 'Genealogical Account' appended to the 'Memoirs' in stating that Glyndwr's mother was descended from Catherine, a daughter of Llywelyn ap Gruffudd, last prince of Gwynedd (d. 1282).[33] There is in fact no evidence that Llywelyn had any daughter other than the Gwenllian who spent her life in a convent in England after the death of her father. Nevertheless, this story became quite widespread from the Tudor period onwards, and probably originated in an attempt to bolster the royal credentials of the Tudor family, since Glyndwr's mother's sister was an ancestor of Henry VII, so that any Gwynedd connection claimed for her inevitably applied to Glyndwr as well.[34] It is clear enough why this connection was advantageous for the patriotic view of Glyndwr, since it represented direct continuity of the struggle for independence, Llywelyn having been the last to claim the title *Princeps Walliae* (Prince of Wales) before Glyndwr himself did so. Focusing on one royal line rather than three has the effect of presenting Glyndwr as a lone individual bravely persevering with a valiant but ultimately hopeless struggle, rather than as one who united an otherwise fragmented nation.

Pennant also drew on various local traditions, either gathered directly from informants by John Lloyd or deriving from earlier manuscript or printed sources. A good deal of folklore seems to have become associated with Glyndwr in the early modern period depicting him as a wily trickster adept at escape, an image which is occasionally evident in Pennant's account.[35] This is sometimes no more than a passing reference not immediately obvious within an otherwise historical narrative, as when Glyndwr's house is suddenly surrounded by the king's forces, 'but he had the good fortune to escape into the woods'.[36] Other pieces of folklore form discrete episodes, such as the story about Hywel Sele of Nannau. This story, in which Hywel is killed by being put into a hollow oak, is known to have existed in essence in the seventeenth century, being associated with an oak tree on the Nannau estate.[37] Pennant, however, was the first to relate the story in full, and it is interesting that he seeks to excuse Glyndwr's horrific deed by presenting it as a punishment for treachery:

> HOWEL SELE of *Nannau* in *Merionethshire*, first cousin to *Owen*, had a harder fate. He likewise was an adherent to the house of *Lancaster*. *Owen* and this chieftain had been long at variance. I have been informed, that the abbot of *Kymmer*, near *Dolgelleu*, in hopes of reconciling them, brought them together, and to all appearance effected his charitable design. While they were walking out, *Owen* observed a doe feeding, and told *Howel*, who was reckoned the best archer of his days, that there was a fine mark for him. *Howel* bent his bow, and, pretending to aim at the doe, suddenly turned and discharged the arrow full at the breast of *Glyndwr*, who fortunately had armour beneath his cloaths, so received no hurt. Enraged at this treachery, he seized on *Sele*, burnt his house, and hurried him away from the place; nor could ever any one learn how he was disposed of: till, forty years after, the skeleton of a large man, such as *Howel*, was discovered in the hollow of a great oak, in which *Owen* was supposed to have immured him in reward of his perfidy. The ruins of the old house are to be seen in *Nannau* park, a mere compost of cinders and ashes.[38]

The phrase 'I have been informed' is perhaps an indication of an oral source for this story, as elsewhere in the *Tours*. John Lloyd might well have had a version of it from Lewis Owen who was descended from a branch of the Nannau family.

The association of a story with features visible in the landscape is a device characteristic of folklore, supposedly guaranteeing veracity. Caves and hillforts were particularly useful in exemplifying Glyndwr's elusiveness.[39] Thus Pennant has the story of Owen hiding in a cave on the Merioneth coast, still called Ogof Owain (Owain's Cave), and the Iron-Age hill fort of Caer Drewyn with its 'vast rampart of stones' overlooking Corwen is cited by him as one of Glyndwr's 'fastnesses among the mountains'.[40]

Evidence from chronicles, local tradition and landscape come together in the encounter near Worcester which forms a high point of Pennant's narrative.

A French army is known to have landed in Pembrokeshire to join forces with Glyndwr in 1405, and according to the fifteenth-century French chronicler Enguerrand de Monstrelet they penetrated into England as far as Worcester, where they faced the king's forces in a stand-off lasting eight days. Monstrelet's account was later followed by Edward Hall, whose interpretation of the events is more favourable to the English side.[41] Pennant's narrative is based closely on Monstrelet, noting differences in Hall. He then gives this vivid description of the situation based on his own fieldwork:

> The camp that *Owen* is supposed to have possessed, is on *Woodbury hill*, in the parish of *Whittley*, exactly nine miles north-west of *Worcester*. It is surrounded with a single foss; and contains near twenty-seven acres. It probably had been an antient *British* post; but was extremely convenient for *Glyndwr*, not only by reason of its strength, but, as *Wales* lay open to him, he had it in his power to retreat among the mountains whenever he found it necessary. The hill is lofty, and of an oblong form. One end is connected with the *Abberley* hills, which, with this of *Woodbury* form a crescent, with the valley, by way of *area*, in the middle. *Henry* lay with his forces on the northern boundary. The brave spirits of each army descended from their posts, and performed deeds of arms in the center between each army. They had a fine slope on each side to rush down to the duel. The *Welsh* especially had a hollowed way, as if formed expressly for the purpose. I surveyed the spot in company with my friend Doctor NASH, and found it answered precisely to the account given by *Monstrelet*.[42]

This is Dr Treadway Russell Nash (1725–1811), whose *Collections for the History of Worcestershire* contains a plan of the camp on Woodbury Hill.[43] Nash's work was published three years after the first volume of Pennant's *Tours*, but Pennant made reference to it in a footnote in the 1784 edition. The site was known locally as 'Owen Glendower's Camp', as attested in the 1695 edition of Camden's *Britannia*.[44] That local tradition does appear to confirm Monstrelet's account, at least in terms of the location of the encounter, and the combination of evidence, together with dating of official records for the month (also cited by Pennant), was sufficient to convince J. E. Lloyd of the authenticity of the episode.[45] However, the most recent authority on Glyndwr, Rees Davies, is of the opinion that Monstrelet's account 'savours of a flight of chivalric literary fancy'.[46] Whatever the historical truth may have been, for Pennant the crucial factor was that landscape corroborated the account in the chronicle, and indeed provided a stage for his imaginative recreation of the scene. And far from being grounds for scepticism, the chivalric element was precisely what appealed to him in this story; not that the Welsh had an opportunity to take the war into England, but that their 'brave spirits' were able to show their worth on equal terms with the best of the French and English.

Monstrelet and Hall disagree (not surprisingly) as to which side backed down first in this stand-off. Pennant notes both opinions, but by ending his narrative with Hall's account of the king chasing the enemy back into Wales he gives the clear impression that that was what actually happened. He then goes on to detail the king's movements between August and October 1405, and explains his purpose in doing so:

> I am thus minute, to shew that *Henry* possessed a strength of body equal to his activity of mind; otherwise he never could have flown with that rapidity from place to place, nor have guarded against enemies so remote as the *Scots* and *Welsh*, at nearly the same period.[47]

Pennant had access to information about the king's campaigns thanks to printed editions of official papers, as already noted. But this is more than just a matter of establishing the sequence of events. Pennant's admiration for the king's energetic campaigning is very evident here, and it is significant that the crisis of 1405 involved threats to England from both Wales and Scotland, not to mention invasion by the French. The eighteenth-century union of England and Scotland is celebrated repeatedly in Pennant's *Tours of Scotland*, and he clearly had considerable sympathy for the efforts to maintain the unity of Britain at this stage in its proto-history.

From the French point of view, the invasion of Wales in 1405 was an incident in the Hundred Years War with England, an attempt to shift the theatre of that war to English soil. Pennant too was conscious of the Hundred Years War as a wider context to Glyndwr's rebellion, viewed very much through the lens of Shakespeare's history plays. The battle of Agincourt (1415) is just over the horizon, and its hero, 'that brave and active prince *Henry of Monmouth*', plays a starring role in the campaign against Glyndwr.[48] The implication is that it was to Glyndwr's credit that it took such a noble warrior to subdue him.

Agincourt has such a compelling historical presence that Pennant interrupts his narrative of David (Dafydd) Gam's attempt to assassinate Glyndwr at his parliament in Machynlleth in 1402 to remind the reader that David Gam's 'unshaken courage' was proved at Agincourt where he gave up his life to save his king on the battlefield, for which he was knighted on his deathbed.[49] This digression, so typical of Pennant's narrative method, has the effect here of setting up two contrary images of David Gam within the same paragraph, as villain of Welsh folklore and as hero of British history.[50] And by introducing Agincourt it also prepares the reader for the passage two pages later where King Henry is said to have given 'strong proof of the high opinion he had of his son *Henry* of *Monmouth*, afterwards king of *England*, at this time only fifteen years of age'.[51] The Glyndwr rebellion is thus placed in perspective as a temporary period of disunity preceding one of the most celebrated triumphs of British history.

Given the loyalist sympathies evident in Pennant's treatment of the king's campaigns, what then of his attitude towards Glyndwr as leader of the rebellion? It has long been recognized that Pennant wrote as a patriot, concerned with the honour of the Welsh people, rather than as a nationalist seeking to promote any political ideals. It should be noted that he appears not to have known of the existence of the two 'Pennal letters' sent by Glyndwr to the king of France in 1406 setting out his vision for a Welsh state including an independent church and two universities, which were a major source of inspiration for Welsh nationalists from the later nineteenth century onwards.[52]

In order to negotiate the contradictions inherent in any celebration of a rebel against the Crown, Pennant adopted two main strategies. One was to refer to Glyndwr primarily as an 'antient Briton' rather than a Welshman, one who 'derived himself from the antient race of *British* princes' and fought for 'the freedom of antient Britons from the galling weight of the Saxon yoke'.[53] This has the effect of distancing the rebellion from contemporary politics, and of placing Glyndwr in the tradition of British resistance to the Romans, as invoked in the very first paragraph of the *Tour*:

> I now speak of my native country, celebrated in our earliest history for its valour and tenacity of its liberty; for the stand it made against the *Romans*; for its slaughter of the legions; and for the subjection of the nation by *Agricola*, who did not dare to attempt his *Caledonian* expedition, and leave behind him unconquered so tremendous an enemy.[54]

The 'now' of the opening phrase surely refers back to Pennant's two published *Tours* of Scotland, as if he is picking up the thread of a conversation begun a decade earlier. And if that is so then the reference to Agricola's Scottish campaign is especially significant; like the English king struggling to maintain a unified kingdom under threat from the Welsh and the Scots, the Romans are depicted as the first to take on the task of conquering and uniting the whole island of Britain. In Pennant's view it was to the honour of the Welsh and Scots that they fought and eventually succumbed to two such illustrious enemies.

The other strategy which Pennant employs in order to play down any political implications of the rebellion is to focus on Glyndwr as an individual detached from his followers. So for instance, an army of eight thousand of Glyndwr's supporters who devastated Grosmont in Monmouthshire before being defeated at

Usk are referred to as 'malecontents'. And interestingly, these Welshmen are denied connection with the ancient British: 'It seems that the *Welsh* forgot the antient spirit of their country; and yielded an easy victory to the enemy'.[55]

Pennant was very conscious of the anti-Welsh prejudice of the English chroniclers, and he was particularly distressed by the claim, first made by Walsingham and repeated by Holinshed, Hayward and Shakespeare, that Welsh women mutilated the genitalia of the English dead after Glyndwr's victory at Bryn Glas.[56] Rather than reject the whole story as propaganda intended to demonize the enemy, as modern historians have done,[57] Pennant sought to counter the slur on the national reputation by citing an earlier source which attributes the deed to a male follower of Glyndwr:

> I wish I could exculpate my countrywomen from this heavy charge. It originates from *Thomas de Walsingham*, an historian who, it must be confessed, wrote within forty years of this event. To his authority I beg leave to oppose that of another antient writer, who ascribes these barbarities to a follower of *Glyndwr*, one *Rees a Gyrch*.[58]

The spectre of mob violence is apparent again in Pennant's treatment of the penal legislation introduced by the English government in response to the rebellion. While he is quite scathing about some of the statutes, such as that forbidding Englishmen from marrying Welsh women, his attitude towards the legislation against public assemblies (*cymhorthau*) is oddly ambiguous. He first quotes a translation from his friend Daines Barrington's *Observations on the Statutes*, 'That no host, rhymer, minstrel, or other vagabond, should presume to assemble or collect together'.[59] He then goes on to explain:

> The word *kymhortha* is mis-spelt from the *Welsh cymmorth*, or the plural *cymmorthau*, assemblies of people to assist a neighbor in any work. Such are very frequently in use at present. There are *cymmorthau* for spinning; for works of husbandry; for coal-carriage. But at this time, these meetings were mere pretences; and their end was the collecting a sufficient number of able-bodied men to make an insurrection. Of such a nature, in old times, were the hunting-matches in *Scotland*. The legislature in that part of *Great Britain* found the evils resulting from them, and at length suppressed them by a law.[60]

Pennant had already commented on the hunting meetings in similar terms in his first *Tour of Scotland*:

> But hunting meetings, among the great men, were often the prelude to rebellion; for under that pretence they collected great bodies of men without suspicion,

which at length occasioned an act of parliament prohibiting such dangerous assemblies.[61]

While the modern *cymorthau* are presented as neighbourly cooperation, the implication of the Scottish parallel is that such gatherings could potentially be a prelude to insurrection, of the kind which had occurred in the corn riots of the 1750s. However, Pennant then goes on to present an alternative view of the medieval *cymorthau*, stressing the inspirational role of the bards in the national struggle for liberty:

> But *cymhorthau* of our countrymen were at this period of a most tremendous nature. They were composed of men the most dreaded by tyrants and usurpers; of BARDS, who animated our nation, by recalling to mind the great exploits of our ancestors, their struggles for liberty, their successful contests with the *Saxon* and *Norman* race for upwards of eight centuries. They rehearsed the cruelty of their antagonists, and did not forget the savage policy of the first *Edward* to their prescribed brethren. They brought before their countrymen the remembrance of antient prophecies. They shewed to them the hero *Glyndwr*, descended from the antient race of our princes; and pronounced, that in him was to be expected the completion of every prediction of our oracular MERLIN.[62]

Temporal distance is again crucial in neutralizing any political impact, and the myth of the massacre of the bards by Edward I was an accepted exemplum of oppression popularized by Gray's 'The Bard' of 1757.[63] Linking to the bardic encomium which heralded the beginning of the rebellion, this passage also has an important function as a patriotic peroration before the account of Glyndwr's last years.

One of the major differences between Pennant's account of Glyndwr's life and that of earlier historians, both English and Welsh, was in the treatment of its end. The received view had been that Owen died a miserable failure, having been forsaken by his own men and forced into hiding. So the 'Memoirs' presents his end:

> After the year 1411, *Owen* was so weakened, his men deserting him, and returning to the king's obedience, that he was forced often to change his quarters and keep less in sight.
>
> A. D. 1415, death put a period to *Owen's* life and misery upon the Eve of *St. Matthew*.[64]

Pennant would have been acutely aware of the need to create a more satisfactory sense of closure to Glyndwr's life, even though his rebellion had undeniably been a failure. This he achieved not by simply denying the received view but by drawing attention to a contemporary document, published by Rymer, which enabled him to provide a fact-based alternative:

> We find that *Glyndwr* maintained his situation for two years longer. In 1415, his affairs bore so respectable an aspect, that the king condescended to enter into a treaty with him; and for that purpose deputed, from the castle at *Porchester*, Sir *Gilbert Talbot*, with full powers to negotiate with *Owen*, and even to offer him and his followers a free pardon, in case they should request it.[65]

This licence, issued by Henry V on 5 July 1415, is indeed a vital piece of evidence regarding the end of Glyndwr's life, although modern historians do not interpret it in such a positive light as Pennant does.[66] In fact it is likely that Henry simply wished to conclude matters in Wales in order to concentrate on more pressing concerns in France. But Pennant had already prepared the way for his interpretation by positing a temporary decline in Owen's fortunes after the defeat at Pwll Melyn in 1405,[67] to which the evidence supposedly relating to the end of his life could be shifted, so that he was able to counter the received view with rational argument before concluding with a defiant rhetorical flourish:

> Both the printed histories, and the manuscript accounts, represent his latter end to have been very miserable; that he wandered from place to place in the habit of a shepherd, in a low and forlorn condition; and was even forced to take shelter in caves and desert places, from the fury of his enemies. This does not wear the face of probability; for, had his situation been so deplorable, majesty would never have condescended to propose terms to such a scourge as *Glyndwr* had been to his kingdom. This retreat, and the distresses he underwent, were probably after the battle of *Pwll Melyn* in 1405, from which he quickly emerged. Death alone deprived *Owen* of the glory of accepting an offered accommodation. The treaty was renewed by the same minister, on the 24th of *February* 1416, with *Meredydd ap Owen*, the son of *Glyndwr*, which it is to be supposed took effect, and peace was restored to *England*, after an indecisive struggle of more than fifteen years. Our chieftain died unsubdued; unfortunate only in foreseeing a second subjugation of his country, after the loss of the great supporter of its independency.[68]

Pennant of course accepts the date of Glyndwr's death given in the 'Memoirs', 20 September, which happens to echo the date on which his rebellion began.[69] The place is said to be the house of one of his daughters, who had married into the families of Scudamore and Monnington, and Pennant invokes 'the tradition of the county of Hereford' in support of the latter option. It is likely that he had that information from the Monnington family, for he states in his account of Glyndwr's children: 'I have had the pleasure of seeing at my house two ladies, owners of the place, direct descendants from the daughter of *Glyndwr*'.[70] Although he had to acknowledge that there is no material evidence for the place of burial, Pennant's phrasing leaves it in no doubt that Glyndwr was indeed interred: 'It is said, that he was buried in the church-yard of *Monnington*; but there is no monument, or any memorial of the spot that contains his remains.'[71]

This is a much more significant point than might appear at first sight. By emphasizing Glyndwr's death and burial Pennant is implicitly denying the myth of the redeemer which grew up around Glyndwr soon after the end of his rebellion.[72] Usually taking the form of the hero sleeping in a cave awaiting the appropriate time to awake and lead his people in their hour of need, this myth, like that of King Arthur, is well attested in both oral form and print in the modern period.[73] One of the factors which fuelled the myth was the fact that Glyndwr's place of burial was unknown. Pennant must have been aware of the myth, and the care which he takes to exclude it from his account can be attributed partly to his enlightened rationalism (like his sceptical treatment of saints' legends), but more specifically to a desire to provide closure not only to the life itself but also to the potentially subversive repercussions of that life. The spirit of Glyndwr is allowed to live on in the patriotic sense of pride in a chieftain who 'died unsubdued', but not in the political sense of any continuation of his struggle for independence. The myth persisted, however, and the irony is that it was Pennant's rehabilitation of Glyndwr which paved the way for the figurative return of the redeemer in political exploitations of the myth from the home rule movement of the late nineteenth century through to the Meibion Glyndŵr arson campaign in the 1980s.[74]

Notes

1 The form of Glyndwr's name adopted here is that used by Pennant himself. The standard modern form of the name is Owain Glyndŵr (or sometimes Glyn Dŵr).

2 The two major works of modern scholarship are J. E. Lloyd, *Owen Glendower* (Oxford: Oxford University Press, 1931), and R. R. Davies, *The Revolt of Owain Glyn Dŵr* (Oxford: Oxford University Press, 1995). On folklore relating to Glyndwr, see Elissa R. Henken, *National Redeemer: Owain Glyndŵr in Welsh Tradition* (Cardiff: University of Wales Press, 1996). All the relevant passages from sources up to Shakespeare's *Henry IV, Part I* are now available in Michael Livingston and John K. Bollard, eds, *Owain Glyndŵr: A Casebook* (Liverpool: Liverpool University Press, 2013). On English attitudes towards Glyndwr, see Alicia Marchant, *The Revolt of Owain Glyndŵr in Medieval English Chronicles* (York: York Medieval Press, 2014).

3 D. Silvan Evans, ed., *Gwaith y Parchedig Evans Evans (Ieuan Brydydd Hir)* (Caernarfon, 1876), 142.

4 Thomas Pennant, *A Tour in Wales. MDCCLXXIII* (London, 1778; 2nd ed. London, 1784), 325 (page references and quotations are from the second edition).

5 Ibid. 394.

6 Among the 'Queries, addressed to the Gentlemen and Clergy of North Britain' published in Appendix III of Pennant's *A Tour in Scotland 1769* (Chester, 1771) is: 'Has the parish given either birth or burial to any man eminent for learning or other remarkable or valuable qualifications?'

7 Idem, *A Tour in Scotland. MDCCLXXII. Part II* (London, 1776), 356–59.

8 Idem, *Tours in Wales*, 3 vols (London: Wilkie and Robinson, etc., 1810), III: 310–92.

9 For a survey of Pennant's network of contacts, see R. Paul Evans, 'Thomas Pennant (1726–1798): "The Father of Cambrian Tourists"', *Welsh History Review* 13 (1986–87): 395–417.

10 On John Lloyd, see R. T. Jenkins, DWB Online *s.n.* Lloyd, John (1733–1793) (accessed 31 July 2014), and R. Paul Evans, 'Reverend John Lloyd of Caerwys (1733–93): Historian, Antiquarian and Genealogist', *Flintshire Historical Society Journal* 31 (1983–84): 109–24.

11 See Paul Evans, 'Thomas Pennant's Writings on North Wales' (unpublished MA thesis, University of Wales, 1985), 27–30.

12 Flintshire Record Office, Hawarden, D/DM/120/3 (quoted in Evans, 'Reverend John Lloyd of Caerwys', 117). Compare the detailed descriptions of the remains of Glyndwr's prison in Pennant, *Tour in Wales*, 387–88.

13 Pennant, *Tour in Wales*, 326.

14 See Lloyd, *Owen Glendower*, 147–48.

15 Lloyd's copies are in Flintshire Record Office, Hawarden, D/E/1394, one said to be transcribed from a manuscript in the hand of John Thomas, which Lloyd collated with 'a MS sent by Mr Price Keeper of the Bodleian Library' (probably Thomas Ellis's text); see Evans, 'Reverend John Lloyd of Caerwys', 117. A second notebook prepared by Lloyd (Flintshire Record Office, Hawarden, D/E/1397) contains another 'Life of Owen Glyndwr', probably his copy of the version in Panton 53 (in the hand of Evan Evans) which attributes the text to Robert Vaughan. J. E. Lloyd, *Owen Glendower*, 148–49, argues that Pennant's source was a copy more closely related to Panton 53 than to the printed 'Memoirs'.

16 Thomas Hearne, ed., *Historia Vitae et Regni Ricardi II* (Oxford, 1729).

17 Pennant refers to the *Scotichronicon* as 'Fordun', i.e. John of Fordun, under whose name it had been published in 1759, but the section relating to Glyndwr was the work of Walter Bower.

18 See Livingston and Bollard, *Casebook*, 200–205. Pennant also used the work of Mademoiselle de Lussan, *Histoire et regne de Charles VI* (Paris, 1753).

19 *The Parliamentary or Constitutional History of England; From the Earliest Times to the Restoration of King Charles the Second [...] By Several Hands. Volume II* (2nd ed. London, 1762).

20 Pennant, *Tour in Wales*, 366.

21 See Evans, 'Reverend John Lloyd of Caerwys', 118–19.

22 Rhys Jones, ed., *Gorchestion Beirdd Cymru* (Amwythig, 1773), 75–79, 81–87.

23 Pennant, *Tour in Wales*, 327, 371. See Henken, *National Redeemer*, 76.

24 Pennant, *Tour in Wales*, 328–30. For text and translation of the poem, see Dafydd Johnston, ed. and trans., *Iolo Goch: Poems* (Llandysul: Gomer Press, 1993), 38–43.

25 Attention was first drawn to Pennant's mistake by Walter Davies (Gwallter Mechain) in *The Cambro-Briton*, I (August, 1820), 458–63 (referring to visits to both sites in 1792). Samuel Lewis, *A Topographical Dictionary of Wales* (London: S. Lewis and Co., 1833), *s.n.*

Llansilin, notes that Glyndyfrdwy is 'where Sycharth has commonly, but erroneously, been supposed to have stood'.

26 Thomas, *History of the Island of Anglesey*, 77. Pennant, *Tour in Wales*, 332. For the original text, see Dafydd Johnston, ed., *Gwaith Lewys Glyn Cothi* (Caerdydd: Gwasg Prifysgol Cymru, 1995), poem 188, lines 25–26.

27 Thomas, *History of the Island of Anglesey*, 84–88.

28 Pennant, *Tour in Wales*, 334–38.

29 On Williams, see DWB Online *s.n.* Williams, Richard (1747–1811) (accessed 31 July 2014). He is thanked in the 'Advertisement' for his 'poetical translations, marked R. W. and for the elegant version of the ode on *Owen Glyndwr*, to which that mark is omitted', Pennant, *Tour in Wales*, [iii]. For a recent edition of the original Welsh text, see Rhiannon Ifans, ed., *Gwaith Gruffudd Llwyd a'r Llygliwiaid Eraill* (Aberystwyth: Canolfan Uwchefrydiau Cymreig a Cheltaidd Prifysgol Cymru, 2000), poem 11.

30 Pennant, *Tour in Wales*, 338.

31 On Glyndwr's genealogy, see Livingston and Bollard, *Casebook*, 425–30.

32 Ibid. 24–29.

33 Thomas, *History of the Island of Anglesey*, 75. The descent from a daughter of Llywelyn is noted in several Welsh genealogical manuscripts of the early modern period, and John Lloyd could have seen it for instance in Peniarth 288 (page 845) owned by Lewis Owen.

34 It is in the Tudor context that Leland notes the connection with Llywelyn in his *Itinerary in Wales*, whilst in the case of Glyndwr it seems to have been promoted initially by the descendants of his daughter Alice who married into the Scudamore family in Herefordshire, see Livingston and Bollard, *Casebook*, 428.

35 See Henken, *National Redeemer*, 88–145.

36 Pennant, *Tour in Wales*, 339; this derives from Panton 53 or the 'Memoirs', see Henken, *National Redeemer*, 100.

37 Ibid. 130–32.

38 Pennant, *Tour in Wales*, 348.

39 See Davies, *Revolt*, 339–40.

40 Pennant, *Tour in Wales*, 371, 339; Henken, *National Redeemer*, 78–79, 124.

41 For the chronicles of Monstrelet and Hall, see Livingston and Bollard, *Casebook*, 200–205, 222–27.

42 Pennant, *Tour in Wales*, 375–76. Monstrelet states that each side took up their positions 'sur une montaigne, et y avoit une grande valée entre les deux ostz' (Livingston and Bollard, *Casebook*, 202).

43 Treadway Russell Nash, *Collections for the History of Worcestershire*, 2 vols (London: John Nichols, 1781–82), II: 465. The plate is reproduced in the extra-illustrated copy of Pennant's *Tours* in NLW, see www.llgc.org.uk/digitalmirror (accessed 30 May 2015).

44 Edmund Gibson, ed., *Camden's Britannia* (London, 1695), 527.

45 Lloyd, *Owen Glendower*, 104–5.

46 Davies, *Revolt*, 194.

47 Pennant, *Tour in Wales*, 377.

48 Ibid. 380.

49 Ibid. 360.

50 On the folklore relating to David Gam, which dates back to the 1650s, see Henken, *National Redeemer*, 98–99.

51 Pennant, *Tour in Wales*, 362.

52 For the two Pennal letters, see Livingston and Bollard, *Casebook*, 120–27.

53 Pennant, *Tour in Wales*, 334.

54 Ibid. 1.

55 Ibid. 369.

56 Walsingham's Latin is translated in Livingston and Bollard, *Casebook*, 161: 'There an atrocity never before heard of was perpetrated, for after the battle Welsh women went to the bodies of the slain, cut off their genitalia, placed the penis of each man in his mouth with the testicles hanging between the teeth and above the chin, and then cut off the dead men's noses and pressed them into their anuses.' For the passages by Holinshed and Hayward, and later Welsh views of the incident, see Henken, *National Redeemer*, 132–33.

57 See R. R. Davies, *Revolt*, 157; Livingston and Bollard, *Casebook*, 454.

58 Pennant, *Tour in Wales*, 352. A footnote cites *Vita Ricardi*, i.e. Thomas Hearne's 1729 edition of the *Historia Vitae et Regni Ricardi II* which John Lloyd consulted in the Mostyn library, see 15n above. The form of the name printed by Hearne (page 178) is actually *Rees à Gytch* with a suspension mark over the *h*. George B. Stow, ed., *Historia Vitae et Regni Ricardi Secundi* (Philadelphia: University of Pennsylvania Press, 1977), reads *Rees a Gythe*, see Livingston and Bollard, *Casebook*, 78–79, where the relevant passage is translated: 'Then a certain Welshman came, of the faction of the said Owain Glyndŵr, by name Rhys ap Gruffudd, who was himself of a fiercer mind even than the others. This one either killed or mutilated or captured all those resisting him.'

59 Daines Barrington, *Observations on the Statutes, Chiefly the More Ancient* (London, 1766).

60 Pennant, *Tour in Wales*, 392.

61 Idem, *Tour in Scotland 1769*, 101.

62 Idem, *Tour in Wales*, 392.

63 See Dafydd Johnston, 'Radical Adaptation: Translations of Welsh Poetry in the 1790s', in *'Footsteps of Liberty and Revolt': Essays on Wales and the French Revolution*, ed. Mary-Ann Constantine and Dafydd Johnston (Cardiff: University of Wales Press, 2013), 169–89 (180–81).

64 Thomas, *History of the Island of Anglesey*, 73. William Wynne, *History of Wales* (London: M. Clark, 1697), 319, gives a similar account of how Glyndwr 'miserably ended his life' (quoted by Henken, *National Redeemer*, 65).

65 Pennant, *Tour in Wales*, 393.

66 For the text of the licence and a similar one giving authority to treat with Glyndwr's son Maredudd, see Livingston and Bollard, *Casebook*, 150–51. For discussion, see Davies, *Revolt*, 326.

67 Pennant, *Tour in Wales*, 370: 'It was at this time that he suffered those distresses which the *English* attribute to the latter part of his life.'

68 Ibid. 394.

69 See discussion by Elissa Henken in Livingston and Bollard, *Casebook*, 579.

70 Pennant, *Tour in Wales*, 331.

71 Ibid. 393–94. On traditions relating to Glyndwr's death, see Gruffydd Aled Williams, *Dyddiau Olaf Owain Glyndŵr* (Talybont: Y Lolfa, 2015).

72 For the earliest evidence of denial of Glyndwr's death, in a Welsh-language chronicle composed soon after 1422, see Livingston and Bollard, *Casebook*, 174–75.

73 See Henken, *National Redeemer*, 64–88.

74 For examples of the use of Glyndwr in Gothic fiction from both these periods, see Jane Aaron, *Welsh Gothic* (Cardiff: University of Wales Press, 2013), 61–63, 130, 163.

Chapter 6

'THE FIRST ANTIQUARY OF HIS COUNTRY': ROBERT RIDDELL'S EXTRA-ILLUSTRATED AND ANNOTATED VOLUMES OF THOMAS PENNANT'S *TOURS IN SCOTLAND*

Ailsa Hutton and Nigel Leask

Recent scholarship has underlined the importance of the practice of 'extra-illustration' or 'grangerization' in constructing antiquarian knowledge in the late Enlightenment: Thomas Pennant's Tours lent themselves particularly well to supplementary visual documentation of this kind. Extra-illustration in the late eighteenth century consisted of refurbishing a published text with an entirely new set of engraved prints, sketches, drawings and watercolours. The additional images were selected according to both personal preference and to the subject of the book. They often took the form of engraved portrait heads, antiquities, maps, topographic views, heraldic displays, newspaper clippings and even letters, and were occasionally supplemented with handwritten notes within the foot and margins of the page. The chosen images were inserted within the book alongside the corresponding text; being either pasted in between pages of printed text, or rebound into separate volumes. Any type of book could lend itself to extra-illustration, from Shakespeare's works to antiquarian, topographical, and travel volumes.[1] The suitability of Pennant's *Tours* through Britain and Europe for this type of personalization and embellishment lay in the fact that they contained numerous references to eminent individuals, antiquities, monuments, landscapes, ethnologies and natural histories. Such wide-ranging subject matter allowed extra-illustrators a broad selection of topics on which to focus their attention.

Recent essays on Pennant's practice of extra-illustration by Paul Evans and Lucy Peltz have focused on *Some Account of London* (1790) and *A Tour in Wales* (1778–83), but Pennant's personal volumes of the *Tours in Scotland*, both 1769 and 1772, held in the National Library of Wales represent another fine example.[2] These volumes contain original drawings by his 'faithful servant' Moses Griffith, as well as works by Paul Sandby, Adam de Cardonnel and John Clerk of Eldin, alongside numerous engravings taken from various publications. A hitherto little-known example by another hand is the extra-illustrated and annotated three-volume set of Pennant's *Tours in Scotland* in the collection of the University of Glasgow Library, made by the Dumfriesshire antiquarian Captain Robert Riddell of Glenriddell (1755–94) in the two or three years before his early death in 1794.[3] Riddell began a correspondence with Pennant around 1790, although the two men never actually met: his volumes, the subject of this essay, are of additional interest to the extent that they represent a Scottish response to Pennant's *Tours*, shedding light on the interrelationship of Welsh, Scottish and English antiquarian networks, and the somewhat vexed construction of a 'British' antiquity in the century following the Act of Union.

Although the *Tours in Scotland* were written by a Welshman and largely aimed at a metropolitan English readership, Riddell's generally positive response to Pennant's text echo that of most Scottish readers, especially in contrast to the furore that accompanied the publication of Dr Johnson's *Journey to the Western Islands of Scotland* (1775) north of the Tweed. Nonetheless, as we will argue, at times Riddell found fault with Pennant's upbeat vision of union, such as his representation of the brutal military suppression of Jacobitism as an unpleasant means that was justified by its end, or his account of aspects of Highland clanship as an atavistic system of rapine and plunder that had been tamed by subjection to British law. Riddell's comments were of course made nearly half a century after Culloden, when the immediate political threat of Jacobitism had been eclipsed by the political events of 1789, a very different climate from the 1770s when Pennant's *Tours* were published. Despite his Foxite Whig politics, and his involvement in the County Reform movement,[4] Riddell, and others like him expressed 'sentimental Jacobitism' as a form of patriotic antiquarianism.[5] In a spirit of collaboration Riddell sought to augment and update any passages in Pennant that were either deficient in information or simply out-of-date. His extra-illustrated volumes thus exemplify the contribution of local experts to national institutions like the Society of Antiquaries of London and of Edinburgh (Riddell was a member of both), as well as the more general Scottish response to Pennant's seminal tour narratives. Our discussion of the making of these extra-illustrated volumes highlights the importance of correspondence, visiting and touring in the working of antiquarian networks; the exchange of drawings and information between agents; the prerogative of accurate visual and verbal description; and the social differences between gentleman antiquaries and their plebeian draughtsmen (recently theorized by Noah Heringman as 'knowledge workers').[6] Although, like Pennant and Francis Grose, Riddell was a skilled topographical artist, evident in the beautiful watercolours included in his manuscript 'Antiquities and Topography of Nithsdale', presented to the Society of Antiquaries in 1793, he seems generally to have delegated the task of visual description to artisanal draughtsmen.[7]

The physical dimensions of Robert Riddell's copy of Pennant's *Tours* exemplify the seemingly haphazard or 'bespoke' nature of antiquarian knowledge, to the extent that the three volumes are assembled from different sets as originally published: Volume I, Pennant's shorter 1769 *Tour*, is the fourth edition of 1776; Volume II, containing the 1772 *Tour* Part I, is the second edition of 1776; and Volume III, containing the 1772 *Tour* Part II, is the first edition of 1776. In addition to containing 41 extra watercolours, wash drawings, etchings, prints and displays, an additional feature of Riddell's volumes not present in Pennant's own extra-illustrated volumes are extensive holograph annotations in ink entered at the foot or in the margins of the pages. These entries are usually signed 'RR' or 'G', standing for 'Robert Riddell' or 'Glenriddell' respectively, and were almost certainly written on different occasions rather than during a single sitting. In a letter of 30 January 1788 to Richard Gough, the Edinburgh antiquary George Paton complained that Riddell's notes on the antiquities of Dumfriesshire, which he had offered to contribute to Gough's new edition of Camden's *Britannia*, were scrawled in ink in the margins of the copy on loan to him, contrary to his express orders. Riddell was clearly a dedicated marginal scribbler: however much this habit may have irritated his colleagues, is it one for which we can be grateful in respect of the volumes discussed here.[8]

The Glenriddell Connection

Robert Riddell of Glenriddell was the scion of an ancient Dumfries family who had been educated at Edinburgh and St Andrews universities. After retiring as a half-pay captain in the 32nd Regiment of Foot in Ireland in 1783 he purchased the estate of Friars' Carse in Dunscore Parish, Nithsdale, Dumfriesshire, and lived there with his wife Elizabeth Kennedy until his death in 1794. Riddell was a Fellow of both the Society of Antiquaries of London and of Edinburgh and of the Manchester Philosophical Society, and Friars' Carse provided him with a base for his antiquarian and literary pursuits, often in collaboration with his Nithside neighbour, the farmer–poet Robert Burns, a connection that has earned Riddell a rather secondary fame in the annals of Scottish literature. In an important article of 1953, Robert Thornton challenged a common view among Burns scholars that Riddell was 'a loud, blustering squire, a hollow and unsubstantial mind', and offered a revised picture of Burns's celebrated patron, drawing upon complimentary references to Riddell in the correspondence between Richard Gough, secretary of the Society of Antiquaries of London, and George Paton, an Edinburgh antiquarian who had in the 1770s served as Thomas Pennant's principal Scottish agent.[9] In 1787 Paton recommended Riddell to Gough as 'a very curious Antiquarian, extremely fond of all our ancient British, Scottish, Roman & other Antiquities and Remains, expert in Genealogical and Heraldic Inquiries', adding that Riddell was anxious to correspond with the Society on the antiquities of his native Dumfriesshire.[10] In 1794 Gough lamented the recent death of the man he described as 'the first antiquary of his country' and noted that 'the letters and communications which passed between us were frequent and interesting'.[11] Gough's high valuation of Riddell is certainly borne out by the latter's publications in *Archaeologia*, the journal of the London Society, and by the list of titles held in the library at Friars' Carse when it was catalogued for auction by Edinburgh bookseller Robert Ross in 1795.[12] As well as being an expert on local topography and genealogy, Riddell was an accomplished musician, publishing two collections of reels and strathspeys, and also collected traditional songs and ballads.[13]

Just as Gough tapped Thomas Pennant for local information about north Wales for his updated, translated edition of Camden's *Britannia* (1789), he also mined Riddell's local expertise for information about Dumfries and Galloway, a region of Scotland the antiquities of which had been neglected until Riddell's articles were published in *Archaeologia*. In 1793, according to a marginal note in Pennant Volume II: 113, Riddell presented 'a Folio MS Volume entitled "The Antiquities and Topography of Nithsdale" – to the antiquarian society of London – in it are about 60 drawings by Capt Grose, and others – a valuable present'. (As mentioned above, this included examples of his own watercolour art.) Regarding his research into Scottish antiquities, 11 volumes of a possible 14 survive dated between 1786 and 1792 and show drawings, prints and descriptions of antiquarian remains drawn and written by Riddell and members of his close antiquarian network. These include Riddell's narratives of three tours in the company of Francis Grose made in May and June 1789 (see below).[14] Riddell was in short part of a local, national and international connection which embraced (by correspondence, visiting, and mutual citation) antiquarians like Lord Hailes, Adam de Cardonnel and George Paton in Edinburgh, the Earl of Buchan at Dryburgh Abbey, Captain George Hutton in Kelso, Thomas Pennant in Wales, Francis Grose and Richard Gough in London, John Nichols, editor of the *Gentleman's Magazine*, and Grimur Jonnson Thorkelin in Copenhagen, and other correspondents in Ireland, France, America and Bengal.

Riddell's most important service to Burns, and to literature, was to introduce him to the English antiquary Captain Francis Grose (1731–1791) during
the latter's first visit to Friars' Carse in 1789, while researching his *Antiquities of
Scotland*, a collection of engraved plates of Scotland's pre-Reformation monuments, with a rather spare letter press.[15] Burns commemorated Grose's first
Scottish tour in his poem 'On the Late Captain Grose's Peregrinations thro'
Scotland, collecting the Antiquities of that Kingdom', with its famous opening
stanza:

> Hear, Land o' Cakes, and brither Scots,
> Frae Maidenkirk to Johnny Groats! –
> If there's a hole in a' your coats,
>   I rede you tent it:
> A chield's amang you, taking notes,
>   And, faith, he'll prent it.[16]

Even more famously, inspired by Burns's account of the popular superstitions
of his native Alloway, Grose commissioned a 'Witch Story' to gloss his plate
of Alloway Kirk, which was the occasion for the composition of the Ayrshire
poet's narrative masterpiece 'Tam o'Shanter'. Less well known perhaps is the
fact that 'Tam o'Shanter' was first published in book form as a monumental
footnote to Grose's *Antiquities of Scotland* in 1791.[17] Perhaps the most impressive
record of Burns's relationship with Riddell, however, are the two presentation
volumes known as the 'Glenriddell Manuscripts', containing transcriptions of
some of the poet's unpublished poetry and prose, prefaced by a generous dedication to Riddell and his wife Elizabeth, 'as a sincere though small tribute of
gratitude for the many many happy hours the Author has spent under their
roof'.[18] Burns intended that these volumes be lodged in his patron's extensive
library in Friars' Carse, but sadly Burns's breach with the Riddell family over
the unfortunate 'Rape of the Sabine Women' incident in December 1793
resulted in his request that the first volume be returned, and Robert Riddell
died before his relations with Burns could be repaired.[19] Riddell's own drawing of 'Robertus Burns, Scotus', the frontispiece of Volume II depicts Riddell's
well-stocked library as a backdrop to the poet's portrait.

The Making of Riddell's Extra-Illustrated Pennant

There is no doubt that Captain Francis Grose was the crucial catalyst in the
genesis and production of Riddell's extra-illustrated volumes of Pennant.
The eccentric and corpulent 'British Antiquarian' was an old friend of
Pennant's, having toured the Isle of Man with him back in 1774, and he had

included drawings by Pennant's artist Moses Griffith in his earlier *Antiquities of England and Wales*.[20] In the summer of 1789, Riddell accompanied Grose on three short antiquarian tours of lowland Scotland: as already mentioned, manuscript accounts of these, written in Riddell's hand, are among the latter's papers held in the Antiquaries Library in Edinburgh.[21] Grose returned to Scotland again in April 1790, a significant date in relation to the Pennant volumes that are the subject of this essay. In a letter to Pennant at Downing from Friars' Carse on 25 August 1790, Grose describes his tour through Perthshire, Fife and Aberdeenshire to complete his researches for *Antiquities of Scotland*, before returning to Riddell at Friars' Carse in late summer, en route to Ireland. It is clear from this letter that at this date Pennant and Riddell were not yet acquainted, for Grose informed Pennant that Riddell 'is a great admirer of your works, & earnestly wishes to be known to you, which he flatters himself with a chance of, as he expects to be in Chester this Winter shold you at any time want information respecting his part of Scotland he will be happy to correspond with you. He is a Gentleman of Honour and Fortune'.[22] Grose was repaying Riddell's hospitality by promoting his entry into Pennant's established and international antiquarian network. This helps us make sense of the letter from Pennant to Riddell, from Downing, dated 31 August 1790 (bound into the front board of *Tour in Scotland*, Volume I), which evidently accompanied a package for Grose: here Pennant advises Riddell that 'should he visit Chester [he] will be very happy to see him here', suggesting that he is replying to Grose's letter dated 25 August.[23] Pennant also requested that if the peripatetic Grose had already departed from Friars' Carse, Riddell forward the enclosed package to him. He had indeed just departed for Ireland in order to begin his 'Antiquities of that Kingdom' – sadly, the corpulent antiquarian (possibly falling victim to his inordinate participation in the same 'Irish hospitality' that had allegedly prevented Pennant from narrating his own earlier tour there) died in Dublin the following year.

Additional light is cast on Francis Grose's role as middleman between the two gentlemen antiquaries Pennant and Riddell, as well as the production of material for the extra-illustrated volumes, by Riddell's annotations in Volume II (Part I of 1772 *Tour*. 88). These describe Grose's second visit to Friars' Carse, which saw the real genesis of the extra-illustrated Pennant volumes:

When Capt Grose came to Friar's Carse in the beginning of May 1790. He brought for my perusal from Mr Pennant the original drawings by Moses Griffiths [*sic*] when attending him on the tour, 1772 and I caused his boy Thomas Cocking copy several of them of which this of Holmshill Tower is one.[24]

This is confirmed by the attribution on the reverse of all of Cocking's draw-ings: 'Copied from Moses Griffith's Drawing, By Thomas Cocking at Friars Carse', followed by the date 'May 1790', and Riddell's signature 'RR'. Thomas Cocking was working hard throughout May, faithfully copying Griffith's 20-year-old originals brought from Wales by Grose, and providing Riddell with the extra-illustrations he needed so that, presumably, these could be returned to Pennant, assuming that the originals were not intended as gifts to convey the Welsh antiquarian's respect for his younger Scottish counterpart.

Within Riddell's volumes there are a total of 41 additional images. They consist of 18 wash and watercolour drawings, 18 etchings and 5 engravings taken from other publications. One of the five engravings was taken from Grose's *Antiquities of Scotland* while another, an engraving after Paul Sandby, came from Robert Heron's *Observations made on a Journey through the Western Counties of Scotland* (1792). Of the 18 drawings that were added to Riddell's volumes one is an unsigned drawing in sepia ink depicting 'Duntulme Castle in Skie' while the other 17 drawings are by Cocking after Griffith. The major-ity of Cocking's drawings are done in pen or pencil and wash and depict a range of subjects such as the castles of Morton, Crookston, Ranza in Arran, Dunvegan, Dunstaffnage, Brodie and MacKinnon in Skye. There are also antiquarian and curious subjects such as fragments of the Ruthwell Cross, a canoe at Kelbane, and a tree at Glenelg church, as well as views of Loch Tarbart and Ardmaddy Bay. That these drawings were made after originals by Griffith neatly demonstrates the practice of lending and copying antiquar-ian and topographic drawings. It also seems probable that possessing draw-ings after the original artists connected with Pennant's *Tours* was desirable and added extra appeal to the customized volumes. Given that Pennant kept a list of Scottish drawings that had been 'presents' from 'Captain Gilpin of Carlisle', Captain Grose, 'Rev Mr Graham, Netherby', 'Mr Jackson, Glasgow', 'Joseph Banks, Esq.', 'Dr Lind' and 'Mr George Paton',[25] antiquarian drawings clearly circulated as part of a 'gift economy', as well as providing a form of exchange currency for reciprocal favours or information rendered.

In Riddell's *Tours* there are also 18 etchings by Adam de Cardonnel (*c*.1746–1820), taken from his *Picturesque Antiquities of Scotland*, which first appeared in London in 1788. Cardonnel illustrated this publication with his own etchings of castles and religious ruins, but the etchings added to the *Tours* by Riddell depict views mainly in southern and central Scotland, such as North Berwick, Dunbar, Linlithgow and Coldstream. Riddell mentions Cardonnel in his annotations in Volume II: 96, where he records, 'In 1788 I carried Mr de Cardonel to Ruthwell'. Riddell's numerous references to Cardonnel confirm the extent of their contact with one another, and another citation written in Volume II: 125 explains, 'I carried both Mr. De Cardonel, and Capt Grose, to visit Morton castle, and they have published views of it in their Respective works […] I have stuck in Mr De Cardonnels etching of that fine ruin.' By looking at the annotations together with the additional illustrations selected and inserted by Riddell, a more precise picture of his antiquarian activities and network in Scotland begins to emerge. Riddell chose to include drawings and etchings by two artists with whom he toured, and who may have made some of their drawings in his presence, seemingly to provide a personal touch to his additions to Pennant's text, and indeed to the Welshman's earlier description of Scotland. We have commented above on Riddell's mission to put the understudied antiquities of Dumfries and Galloway 'on the map', so it is no surprise that his choice of illustrations (and indeed marginal comments) focuses on the antiquities of southern Scotland and the Scottish Borders, rather than on the Highlands and Islands where his knowledge was on weaker foundations.

Pennant's eight-week 'Voyage to the Hebrides' ended on 12 August 1772, when the *Lady Frederic Campbell* anchored in Ardmaddy Bay, near Seil Island in Argyll. Within Riddell's volumes is a beautiful watercolour copied by Cocking, pasted into the inside board of Volume II. The watercolour's caption reads 'Ardmaddie Bay' and points the reader to page 420 of the same volume where Riddell explains that the drawing was 'copied by Thos Cocking from Moses Griffiths Drawing at Friars Carse May 13 1790' (Fig. 6.1). The original drawing of Ardmaddy Bay by Griffith which Cocking copied can be located within Pennant's own extra-illustrated volumes of the Scottish *Tours* found in the National Library of Wales. This original drawing is unsigned and undated, as with many of the images contained within Pennant's extra-illustrated volumes, but upon comparison of the two images the connection is made clear. Like his precursor Moses Griffith, Cocking was regarded as a copy-artist and part of his role was to copy multiple drawings to be distributed within Grose's publications and among members of his social circle. Grose describes drawings made by Cocking and published in his *Antiquities of Scotland* as being 'drawn, under my inspection, by my servant, Thomas Cocking, a young man who promises to make an accurate draughtsman'.[26] Cocking's part within both Grose and Riddell's picture-making also highlights the value of copies within antiquarian networks as antiquaries borrowed and exchanged drawings to be copied, displayed, studied and appreciated as documentary records in their own right.

Riddell's Marginal Annotations

Riddell's marginal annotations provide a valuable verbal supplement to the copious extra-illustrations discussed in the previous section. His comments are

A TOUR

Cross the *Efa*, through a ford with a bottom of solid rock, having on one side the water precipitating itself down a precipice forming a small cataract, which would afford a scene not the most agreeable to a timid mind. The water too was of the most crystalline, or colorless clearness, no stream I have ever seen being comparable; so that persons who ford this river are often led into difficulties, by being deceived as to its depth, for the great transparency gives it an unreal shallowness.

This river is inhabited by trouts, parrs, lochs, minnows, eels and lampries; and what is singular, the chub, which with us loves only the deep and full waters bounded by clayey banks.

Hol-house.

On the opposite eminence see *Hol-house*, a defensible tower like that at *Kirk andrews*, and one of the seats of the famous *Johnny Armstrong*, laird of *Gilnockie*, the most popular and potent thief of his time, and who laid the whole *English* borders under contribution, but never injured any of his own countrymen. He always was attended with twenty-four gentlemen well mounted: and when *James* V. went his progress in 1528, expressly to free the country from marauders of this kind, *Gilnockie* appeared before him with thirty-six persons in his train*, most gorgeously apparelled; and humbled so richly dressed, that the king said *what wants that knave that a king should have?* his majesty ordered him and his followers to immediate execution, in spite of the great offers *Gilnockie* made; who finding all application for favor, vain, he according to the old ballad, boldly told the king,

Johnny Armstrong.

* *Lindsay*, 147.

To

Figure 6.1 *A Tour in Scotland, and Voyage to the Hebrides, MDCCLXXII, Part I*, 2nd ed. (London, 1776), with extra-illustration of Holhouse Tower, by Thomas Cocking after Moses Griffith, and with marginal notes by Robert Riddell of Glenriddell.

on the whole wide-ranging and often refer to his own personal experience of specific sites and antiquarian objects: unlike Pennant, however, Riddell appears to have had little interest in natural history. As is evident from the following brief survey of some of the highlights of Riddell's annotations to the three volumes, he makes frequent reference to objects or books in his own collection, as well as inserting bibliographical cross-references to works by William Camden, Alexander Gordon, Martin Martin, William Roy and Horace Walpole, or to prints by Hearne and Byrne.[27] The first volume (the 1769 *Tour*) contains few annotations until we reach the Borders and North Berwick. Against Pennant's account of the battle of Dunbar (1650) when Cromwell defeated the Scots army (I: 57), Riddell has written 'I lately procured the very fine Medal executed by Simon which Cromwell caused to be struck on the victory at Dunbar.' This exemplifies Riddell's zeal as a collector of precious antiquities, as well as the connoisseur's pride in detailing objects in his personal collection, which perhaps more than anything else distinguishes the practice of eighteenth-century antiquarianism from the discipline of modern archaeology. It is also evident in his note on Pennant's gloss on his illustrations of 'A small brazen Hermes or Terminus [...] ornamented with festoons and fruits', and 'a brass head of Jupiter, both observed in the house of Revd Graham at Netherby', from Part I of the 1772 *Tour* (II: 84): 'This fine figure is now in my collection. This Jupiter I also have'. There is no doubt that Riddell's acquisitive urge was that of a bona fide connoisseur, but as Diderot satirically represented the antiquarian disposition in his definition of a *curieux* in the *Encyclopedie*:

> Not every one who indulges in [curiosity] is a connoisseur; and that is why enthusiasts are so often figures of fun [...] However, curiosity, this desire for possession, which is almost always without limits, is almost always deleterious to one's pocket.[28]

Returning to the 1769 *Tour*, Riddell's marginal annotations on Pennant's account of Edinburgh reflect his national pride in the visual splendors of late Enlightenment Scotland. Writing 20 years after the Welshman's account of the capital, when James Craig's New Town had transformed the urban landscape, Riddell writes (I: 68): 'George Square is now completed and is indeed a very elegant, airy, and uniform place [...] when the university is built upon Mr Adam's plan the city will be worth Mr Pennant's revisiting.' Ekphrastic descriptions of paintings observed in Scotland's palaces and stately homes were a recurrent feature in eighteenth-century tour accounts, and Pennant's was no exception. Riddell's marginalia has much to say about paintings (especially some of the historical portraits) that Pennant had missed (e.g., I: 60), and he records a 1789 visit to the Edinburgh studio of David Allan to view the latter's history paintings of episodes from the life of Mary Queen of Scots, now unfortunately lost (I: 80).[29] Riddell also supplemented Pennant's topographical descriptions with references to contemporary landscape painting, as for example at I: 77, where he comments on his description of the prospect of the Firth of Forth: 'Mr Miller of Dalswinton employ'd Mr Nasmith [*sic*] the painter to make a picture of this fine view which I have seen.' Again, as so often, Riddell adds a personal note of witness to his observation.

Pennant's interest in Scottish popular culture, religion and superstitions is one of the most striking features of his *Tours*, and clearly resonated with Grose (author of *A Provincial Glossary, With a Collection of Local Proverbs and Popular Superstitions*, 1787), as well as with Riddell.[30] This is evident in Riddell's annotation in the margins of Pennant's account of a Presbyterian festal Eucharist or 'Holy Fair', observed near Kenmore in Perthshire, a popular communion sacrament more famously described in Burns's poem 'The Holy Fair'. Pennant commented, by means of an allusion to Swift's *Tale of a Tub*, on the coarse and licentious handling of the sacraments by the celebrants: 'before the day is at an end, fighting and other indecencies ensue. It has often been made a season for debauchery; and to this day, Jack cannot always be persuaded to eat his meat like a Christian' (I: 101). Riddell has written here: 'Vide: Burns's Holy Fair – a poem that lashes this shameful abuse RR'. (In fact Burns had concluded in a rather less censorious manner: 'There's some are fou o' *love divine*; / There's some are fou o' *brandy*; / An' monie jobs that day begin, / May end in *Houghmagandie* / Some ither day.')[31] Critics have been divided as to Burns's attitude to the popular Calvinism anatomized here, but Riddell doesn't seem to harbour any doubts concerning the critical intent of his friend's poem 'The Holy Fair'.[32] This is unfortunately the only marginal note in which Riddell mentions his friend Burns directly, but it is likely that the poet, who had made his own Highland tour in 1787, had read the three volumes of Pennant's *Tours* in Riddell's library, if not at some earlier date.[33]

In a passage in the 1769 *Tour* interrupting his account of Rannoch, and the forfeited estates of the Jacobite poet Robertson of Struan, Pennant dilated on a number of Highland superstitions, which he felt might be useful in 'teaching the unshackled and enlightened mind the difference between the pure ceremonies of religion, and the wild and anile flights of superstition' (I: 95). These included belief in spectres, unlucky days (*La Sheachanna na bleanagh*), second sight, the Gaelic Beltane Mayday ceremony and the *coranich* or funeral song, which Pennant had witnessed in Ireland and described with reference to Camden (I: 100). Pennant is particularly interested in antiquarian collectibles to which were attributed magical properties, such as 'elf-shots' (ancient flint arrowheads allegedly with powers to harm or cure cattle), 'crystal gems' and 'adder stones', which from time immemorial had been associated with good fortune and the curing of diseases (I: 101). Riddell's marginal note to this passage is especially interesting, to the extent that it glosses Pennant's observation with reference to a comparable object, which he had received as an heirloom, rather than as an antiquarian collectible:

> I am in possession of a crystal triangular spheroidical gem, sett in silver, which has been long in my family. It is called the Connach stone, and the manner of using it is as follows. Take a large pail of water, and lay across the mouth, a stick north and south, then suspend the Gem with a thread from the pole into the water where it may hang 24 hours then give the water to the cattle who are diseased. G: [I: 115]

Pennant returns to this theme in his 'Voyage to the Hebrides', where Macdonald of Kingsburgh presents him with a *glain naidr*, known vulgarly as a 'Druidical bead', taking its name from the figure of coiled serpents marked on its glass surface. Drawing for once upon his local Welsh knowledge, he writes: 'The common people in Wales and in Scotland retain the same superstitions relating to it as the ancients, and call it by the name of *Serpent-Stone*' (II: 342). He cites the poet Mason's description of Druidical power invested in 'the Druid wand, and the potent adder-stone'. Pennant describes the popular superstitions still attached to such beads, such as curing the bite of the adder, or easing the pangs of childbirth: 'So difficult it is to root out follies that have the sanction of antiquity', he adds, with an enlightened flourish (II: 344). Interestingly, Pennant's description of the *glain naidr* doesn't elicit any marginal annotation from Riddell, who provides little commentary on the Hebridean and Highland sections, presumably because they were beyond the area of his geographical expertise.

As mentioned above, Riddell sometimes bridled at Pennant's Unionist prejudices. For example, he corrects Pennant's scepticism about the authenticity of James Macpherson's *Ossian* poems, remarking that 'Mr Pennant need not have spoke of this in a doubtful manner as it is now beyond dispute' (I: 215). (This was in fact far from being the case, and even the 'official' enquiry by the Highland Society in 1803 returned a somewhat inconclusive verdict.) However, perhaps the strongest reaction from Riddell is his comment on Pennant's glossing over the atrocities committed by the Duke of Cumberland's army after the battle of Culloden 1746 (I: 175). Commending Duncan Forbes's unsuccessful attempts to obtain clemency for the rebels ('to sheath, after victory, the unsatiated sword'), Pennant had written: 'But let a veil be flung over a few excesses consequential of a day, productive of so much benefit to the united kingdoms.' Here an indignant Riddell wrote in the margin: 'Not a few excesses, but a most shameful and unfeeling Carnage of poor wretches, who were slain in cold blood; an officer was Broke for his refewsing [*sic*] to execute one of these nocturnal slaughters of sleeping victims.' Slightly qualifying his indictment of Hanoverian cruelty, he added: 'But when George 2nd read the proceedings tho in compliance to his son, he confirmed the sentence, Yet to his honour he gave the honest officer ascension.'

Although Riddell's father Walter was apparently one of two hostages taken by the Jacobite army in 1745,[34] as noted above, Riddell's marginal comments reveal his 'sentimental Jacobitism', especially in his panegyric on Donald Cameron of Lochiel, one of the leading Jacobite commanders in the '45, the 'Memoirs' of whose celebrated grandfather Ewan Cameron appeared as an Appendix to Pennant's *Tour* (I: 203). Elaborating on Pennant's representation of the younger Lochiel's 'well-intending heart, overpowered by the unhappy prejudices of education', Riddell added a personal recollection: 'I have often heard my Grandfather talk of Locheil with raptures, he said he equaled an old virtuous Roman for Greatness of Soul and high Ideas of Honour – G.' Riddell's other explicit criticism of Pennant occurs when the outlawed Clan MacGregor are dismissed as 'a murderous clan, infamous for excesses of all kinds', proscribed for their massacre of the Colquhouns in 1602, and whose descendants, marked off by 'the redness of their hair', have retained 'the mischievous dispositions of their ancestors' (I: 223). 'This is the first ill-natured remark that Pennant has made in this volume', writes Riddell, before representing the MacGregors as victims of Campbell expansionism in the blood-soaked clan feuds of the seventeenth century (I: 244). The latter clan, he insisted, 'had the address to get them [the MacGregors] proscribed from attempting to recover their estates from the Campbells, who had unjustly got them wrested from the MacGregors. Taymouth and Kilchurn Castles, were the habitations of the Laird of the MacGregor. G.' (I: 244–5). This was of course far from being the standard Whig view of Highland history.

The latter parts of Riddell's copy of the *Tour* 1769, describing Pennant's return journey through the Borders and Northern England, are particularly heavily annotated, as are the first two hundred pages of Riddell's second volume, describing Pennant's 1772 *Tour* through the Lowlands, in contrast to the sparse comments in the 'Voyage to the Hebrides' that makes up the greater part of that volume. One interesting gloss that appears early in Volume II is Riddell's account of his role in the antiquarian study of the Ruthwell Cross, near Annan, in Dumfriesshire, an eighth-century Anglian cross of great importance, with runic inscriptions, smashed up by order of the General Assembly in 1644, as noted by Pennant. Riddell adds in his marginal commentary:

> In 1788 I carried Mr de Cardonnel to Ruthwell, and procuring a number of assistants had all the fragments of this celebrated monument collected which he

drew in a very masterly and accurate manner, and his drawings were ingraved by the Antiquarian society of London, and published in their Vetusta Monumenta, Vol 2, 1789. So this very curious antiquity is now preserved to the learned to Examine [II: 97–8].

A copy of Cardonnel's drawings is bound in here by Riddell. Recovering and describing the fragments of the Ruthwell Cross seems to have been one of Riddell's most important services as an antiquarian, involving collaboration with George Paton and Richard Gough, as well as with Cardonnel. Robert Thornton describes how Cardonnel's sketches of the Ruthwell Cross were shipwrecked in the *Duchess of Buccleugh* en route from Leith to London, but because they were contained in a sealed tin box, they were preserved from destruction. Grimur Thorkelin received copies, and a learned article was published on the Cross in Copenhagen in 1837.[35]

The third and final volume, detailing the remainder of Pennant's 1772 tour from Argyll, across Perthshire to Fife and Angus, before his return trip south to the Borders and England, contains no annotations, until we reach Brechin, Angus; thereafter frequent marginal comments resume. We'll finish by discussing Riddell's marginalia to Pennant's account of the University of St Andrews, which the Welsh traveller had visited in the autumn of 1772, when Riddell was himself a student attending the university. The two men didn't meet on that occasion (Riddell was of course 30 years Pennant's junior), and despite their shared membership of the Society of Antiquaries, and Pennant's 1790 invitation to Riddell to visit him at Downing, there is no evidence that they ever did meet in person. So this particular marginalia represents a fortuitous personal connection between the Welsh and Scottish antiquaries who shared so many common interests. Compared to Dr Johnson's account of St Andrews as 'a university declining, a college alienated, and a church profaned and mouldering into dust',[36] Pennant was highly complimentary. He described the university as inexpensive, the board excellent, and its professors 'indefatigable in their attention to the instruction of the students, and to that essential article, their morals […] no little irregularity can be committed, but it is instantly discovered and checked'. In the bottom margin, however, Riddell takes issue with this judgement, as an alumnus of Scotland's most ancient university:

> This I know was not true; in 1772 I was at the university of St Andrews and found it more expensive than Edinburgh. The Professors (except the worthy Principal Tullidaff) were at no pains, nor took the slightest care of any students, who did not reside in their house. R[obert].Watson the Prof of Rhetorick had 5 Boarders who each paid £100 per annum. I sometimes dined at the college

table – which was shamefully bad – every kind of dissipation was carried on openly – and never checked by any professor, except when it reached the ear of Principal Tullidaff of whom every Professor stood in awe [III: 198].[37]

This marginalia also tells us that Robert Riddell was tutored by Professor Robert Watson at St Andrews, recently acknowledged (alongside Adam Smith and Hugh Blair) as one of the founding fathers of the modern discipline of university English.[38]

In this essay we have examined Riddell's extra-illustrations to his volumes of Pennant's *Tours in Scotland*, as well as the annotations that complement them, in their relation to the original text. We've shown Riddell embodying the distinctively eighteenth-century intellectual persona of the antiquary–scholar, obsessed by collecting, drawing, corresponding, and library building, many of his other books quite possibly customized along the lines of the examples discussed in this essay. In this respect, he was following the example of his hero Thomas Pennant, and thereby participating in a significant Cambrian–Caledonian intellectual collaboration. Riddell's extra-illustrated *Tours* are a good example of how Pennant's works could be customized and personalized accordingly, and they also present a significant response to Pennant's account from a native Scot. It is hard to imagine that many other items in Riddell's library were as richly annotated and grangerized as his magnificent three-volume set of Pennant's *Tours of Scotland*, although sadly, as we've seen, that library was dispersed shortly after his premature death in 1794.

Notes

1 Lucy Peltz, 'The Extra-Illustration of London: Leisure, Sociability and the Antiquarian City in the Late Eighteenth Century' (unpublished PhD thesis, University of Manchester, 1997), Appendix II.
2 See Ibid.; idem, 'The Extra-Illustration of London: The Gendered Spaces and Practices of Antiquarianism in the Late Eighteenth Century', in *Producing the Past: Aspects of Antiquarian Culture and Practice, 1700–1850*, ed. Martin Myrone et al. (Aldershot: Ashgate, 1999), 115–34; R. Paul Evans, 'Richard Bull and Thomas Pennant: Virtuosi in the Art of Grangerisation or Extra-Illustration', *National Library of Wales Journal* 30, no. 3 (1998): 269–94.
3 Riddell's *Tours* were purchased by David Murray and donated to the University of Glasgow Library in 1928.
4 In 1792 Riddell contributed essays on county reform in the radical *Edinburgh Gazetteer*, under the pseudonym of 'Cato'. See Robert Burns, 'On Glenriddell's Fox Breaking his Chains', in *Poems and Songs of Robert Burns*, ed. James Kinsley, 3 vols (Oxford: Clarendon Press, 1968), poem 527; Nigel Leask, *Robert Burns and Pastoral: Poetry and Improvement in Late Eighteenth-Century Scotland* (Oxford: Oxford University Press, 2010), 260.
5 Anti-Catholic and anti-Jacobite laws and the ban on tartan were lifted in 1782 two years before the restoration of the forfeited estates was begun. See Colin Kidd, 'The Rehabilitation of Scottish Jacobitism', *The Scottish Historical Review* 77, no. 203 (1998): 58–76 (62).

6 The term is Noah Heringman's, *Sciences of Antiquity: Romantic Antiquarianism, Natural History, and Knowledge Work* (Oxford: Oxford University Press, 2013).

7 An example of Riddell's skill as a watercolourist ('Caerlavarock Castle') is published in David Gaimster et al., eds, *Making History: Antiquaries in Britain, 1707–2007* (London: Royal Academy of Arts, 2007), 70.

8 Quoted by Robert Thornton, 'Robert Riddell: Antiquary', *Burns Chronicle*, 3rd series, II (1953): 44–67 (55). Riddell's marginal notes in Richard Gough's *British Topography* are discussed in James Holloway and Lindsay Errington, *The Discovery of Scotland: The Appreciation of Scottish Scenery through Two Centuries of Painting* (Edinburgh: National Gallery of Scotland, 1978), 73.

9 Thornton, 'Robert Riddell: Antiquary', 44. The correspondence is held in NLS.

10 Ibid. 48.

11 Ibid. 64.

12 *A Catalogue of Curious and Valuable Books on Antiquities, Collected at Great Expense, by an eminent Antiquary, late Deceased* (Edinburgh 1795). The auction was held at Robert Ross's auction room, Edinburgh on 2 March 1795. Thanks to Robert Maclean for locating this reference.

13 The 'Glenriddell MS', 1791, is volume 9 of Riddell's MS Collection and contains his collection of songs and ballads: see 14n. It was an important source for Sir Walter Scott's *Minstrelsy of the Scottish Border* (1802–3). Riddell also published *A Collection of Scotch, Galwegian and Border Tunes* (1794). See David Johnson, *Music and Society in Lowland Scotland in the 18th Century* (London: Oxford University Press, 1972), 118.

14 Robert Riddell's MS Collection (11 vols) containing accounts of history, families, and antiquities of Dumfriesshire, Galloway etc., 1786–92: National Museums Scotland Library, Appendix MSS. 581–91. See also Holloway and Errington, *Discovery of Scotland*, 57–59.

15 J. H. Farrant, ODNB Online *s.n.* Grose, Francis (*bap.* 1731, *d.* 1791) (accessed 1 February 2014).

16 Burns, *Poems and Songs of Robert Burns*, ed. Kinsley, poem 275, lines 1–6. First published under the title 'Address to the People of Scotland' in the *Edinburgh Evening Courant*, 11 August 1789, and collected in the Edinburgh edition of Burns's *Poems* in 1793.

17 Leask, *Robert Burns and Pastoral*, 256–75.

18 For a fully annotated text, see *Burns's Commonplace Books, Tour Journals and Miscellaneous Prose*, ed. Nigel Leask (Oxford: Oxford University Press, 2014), 175.

19 For an account of this often misunderstood event, as well as much information on the Riddell circle, see Angus Macnaughten, *Burns's Mrs Riddell* (Peterhead and Hythe: Volturna Press, 1975).

20 Evans, 'Richard Bull and Thomas Pennant', 290 n. 48.

21 See 13n above, Robert Riddell MS Collection, Vol. 8, 131–289.

22 NLW, TP244/2, Grose to Pennant, 25 August 1790 [for 1780]. My thanks to Mary-Ann Constantine for finding this letter.

23 Inserted by the former owner of the University of Glasgow copy of Riddell's Pennant, David Murray, who adds a note explaining Riddell's annotation along with the letter, which is pasted in the front cover of the first volume: University of Glasgow Library, Special Collections, Mu6-d.9. The letter is marked in pencil 'acc 2426'.

24 In Riddell's 'Journal of a Tour in Scotland in 1789' (2nd tour, 10 June 1789), 207, Riddell includes eight prints from 'Pennant's octavo edition, published before the

quarto' – these were given to Riddell by Grose in May 1790. (It must have been the 1769 Tour, containing 18 plates; 21 new plates were added to quarto edition, published 1774).

25 NLW, 2530A, Thomas Pennant, 'Presents of Drawing', undated.

26 Francis Grose, *The Antiquities of Scotland* (London: printed for S. Hooper, 1789–91), I: xxi.

27 For example, references to Hearne and Byrne, see II: 31, 51, 65, 186; III: 223. For references to Cordiner, see I: 200, 353–58. For reference to Slezer, see III: 218. For references to Cardonnel, see II: 63; III: 208, 269. For references to Roy, see I: 168, 262; II: 79, 102. For reference to Walpole, see I: 85.

28 Quoted in Krzysztof Pomian, *Collectors and Curiosities: Paris and Venice, 1500–1800*, trans. Elizabeth Wiles-Porter (Cambridge: Polity Press, 1990), 132.

29 As noted by Basil Skinner in his catalogue to the 1973 exhibition '*The Indefatigable Mr Allan': The Perceptive and Varied Work of David Allan, 1744–1796, Scotland's First Genre Painter* (Edinburgh: Scottish Arts Council, 1973), 22.

30 For an excellent overview, see Marilyn Butler, 'Antiquarianism (Popular)', in *An Oxford Companion to the Romantic Age: British Culture 1776–1832*, ed. Iain McCalman (Oxford: Oxford University Press, 1999), 328–38.

31 Burns, *Poems and Songs of Robert Burns*, ed. Kinsley, poem 70, lines 239–43.

32 See Leask, *Robert Burns and Pastoral*, 191–97.

33 *Robert Burns's Commonplace Books*, ed. Leask, 374 n. 231.

34 See James A. Mackay, ODNB Online *s.n.* Riddell, Robert, of Glenriddell (*bap.* 1755, *d.* 1794); Fitzroy Maclean, *Bonnie Prince Charlie* (Edinburgh: Canongate, 1988), 139.

35 Thornton, 'Robert Riddell: Antiquary', 57.

36 Samuel Johnson, *Journey to the Western Islands of Scotland*, ed. R. W. Chapman (Oxford: Oxford University Press, 1970), 8.

37 Principal Tullideph is represented as the fearsome 'Pauly Tam' in Robert Fergusson's 'Elegy on John Hogg' – the famous Edinburgh poet, Burns's 'elder brother in misfortune', had also been a student at St Andrews between 1764–68. This marginalia is wrongly attributed to Francis Grose by Matthew McDiarmid in *Poems of Robert Fergusson*, ed. Matthew P. McDiarmid, 2 vols (Edinburgh: Scottish Texts Society, 1954–56), I: 15.

38 According to Robert Crawford, *Scottish Invention of English Literature* (Cambridge: Cambridge University Press, 1998), 10–11, Watson was 'the second professor of the new art of polite letters' at any British university to hold a chair whose title contained the term 'Rhetorick', following Adam Smith's appointment at the University of Glasgow in 1751.

Chapter 7

'A GALAXY OF THE BLENDED LIGHTS': THE RECEPTION OF THOMAS PENNANT

Elizabeth Edwards

In a preface to the first of the six volumes of domestic tours he published between 1798 and 1800 as *The British Tourists; or Traveller's Pocket Companion*, William Fordyce Mavor began with a dry comment on the fact that the status and popularity of the Grand Tour had left Britain's home territories comparatively little known:

> It was long a reflection on the national taste and judgment, that our people of fashion knew something, from ocular demonstration, of the general appearance of every country in Europe, except their own.[1]

This was a commonly held view in the period, but for a Scottish Whig like Mavor, the point was that the outstanding features of modern Britain were all on show in the home tour. '[I]n whatever light we regard the British Islands;' he observed:

> whether as the cradle of liberty, the mother of arts and sciences, the nurse of manufactures, the mistress of the sea; or whether we contemplate their genial soil, their mild climate, their various natural and artificial curiosities, we shall find no equal extent of territory, on the face of the globe, of more importance, or containing more attractions [...]. (v–vi)

The project Mavor was introducing was his selection of the domestic tours published in the last third of the eighteenth century, which, at least on his terms, displayed the genre at its best. This work was, he stated, oriented towards patriotism and benevolence, framed in terms of 'utility and propriety' as a contribution to 'to the public good' (viii). The excerpted, collected tours were intended to detail British improvement and advancement – subjects that Mavor, who was from a modest background, was keen should circulate to the widest possible audience. Modern tours were, he noted, the preserve of the wealthy due to their cost, but he optimistically imagined *The British Tourists* as a way of putting them 'within the reach of every class of his fellow subjects' (ix). In tandem with the levelling sentiment of this comment, the project was one of enlightenment and education, and Mavor used metaphors of illumination to say more about the public benefit of collecting and abridging tours. The six-volume series 'collects, into one focus, the scattered rays of information; or, rather, it forms a galaxy of the blended lights, which distinguished modern tourists have thrown on the British Isles' (xi).

Front and centre in this galaxy of modern tourists was Thomas Pennant. Selections from the Scottish tours he made between 1769 and 1772 fill the first volume of the series; his *Journey from Chester to London* (1782) and *Some Account of London* (1790) appear in volumes three and six respectively. Pennant's 1769 tour was the earliest published travelogue included by Mavor. But the distinction of being placed first in this 'best of' collection of recent tours was not just chronological happenstance. From their first publication, Pennant's Scottish travels – quickly followed by Samuel Johnson's in 1775 – established the taste for Scottish tourism. As Mavor put it, Pennant 'paved the way to that general taste for home travels': 'to this gentleman we are indebted for the earliest tours in Scotland' (1). While he also included Johnson's *Journey to the Western Islands of Scotland* in the second volume of *The British Tourists*, Mavor strongly resisted its portrayal of Scotland – though not without enjoying the irony that Johnson's 'petulant remarks on Scotland, roused the pride of the natives into exertion, waked every generous passion in their breasts to excel; and, perhaps without intending it, proved himself one of their best friends' (vii–viii). I will consider questions of patriotism and home tourism in the final section of this chapter, but by comparison with Johnson, Pennant was for Mavor the ideal Enlightenment traveller, touring Scotland with 'a party formed and equipped in the manner of an eighteenth-century scientific expedition', including a botanist (Revd John Lightfoot) and a Gaelic scholar (Revd John Stuart), as well as his resident artist, Moses Griffith.[2]

Starting from this sense of Pennant as one of the best-known and best-regarded cultural topographers of late eighteenth-century Britain, this chapter explores the reception of his work from the 1770s to the 1820s. Dealing with the extensive reception history of several different texts over half a century – even the *Tours* of Scotland and Wales cannot be viewed as a single unit in any straightforward sense – is, however, no straightforward matter. In this chapter, I suggest three broad divisions within that history. The first section of the chapter surveys uses of Pennant's Welsh and Scottish *Tours* in a wide-ranging set of texts, from creative to historical to scientific and agricultural. These uses of his work – some fleeting, others much more sustained – illustrate his place within provincial and national conceptions of Enlightenment. The second section focuses on literary responses to Pennant from visitors to 1790s Wales: references to the *Tour in Wales* by William Sotheby and Anna Seward confirm his central position in Romantic-period perceptions of the Welsh past. The final section brings together the previous two to suggest that posthumous uses of Pennant show him increasingly perceived as part of a wider Welsh Enlightenment–part of a national archive of texts that offers a platform for continued development, rather than a nostalgic lens on the past. In conclusion, I suggest that his work may be read as part of a larger cultural nationalism that, in contrast to Katie Trumpener's nostalgic, often melancholy model of bardic nationalism, marked by grief and loss, looks forward rather than back. Characterized by discovery and dialogue (often of a scholarly nature, at times heavily footnoted), this strand of Pennant's reception takes an Enlightenment–antiquarian taxonomy of what survives in Wales as its theme, and as the means to future national progress.

Agreeable Miscellany: Responses to the Welsh and Scottish *Tours*

In 1801, the *British Critic* reviewed Thomas Pennant's two-volume *A Journey from London to the Isle of Wight* (1801) with a burst of praise for his mixed style:

> Mr. Pennant's productions of this kind have afforded so agreeable a miscellany, such a happy mixture of entertaining and instructive matter, diversified by anecdote, history, and above all by the display of his talents as a naturalist, that they have constantly been received by the public with eagerness, and have obtained a place in most well-chosen libraries.[3]

Pennant's many works had built up a sizeable reception history by the time the *British Critic* offered this view. He was a reference point for all the major travel writers who follow him: in the case of Wales, his *Tours* appear in Richard Warner's *A Walk through Wales, in August 1797* (1798) and *Second Walk through Wales [...] in August and September 1798* (Bath, 1799) – which, on describing Chester simply concedes that Pennant has said it all – in William Bingley's *A Tour round North Wales, Performed during the summer of 1798* (1800), and John Evans's *A Tour through part of North Wales, in the year 1798* (1800). Pennant's career is usually viewed in the aggregate, as in the *British Critic's* 1801 review, but the reception of his Welsh and Scottish *Tours* take on different shapes due to their dates of publication, which partly determine the rate at which they pass into wider usage and start being used as sources for guidebooks, authoritative accounts of places, artefacts, buildings, local manners and customs. There are few references to Pennant's Welsh *Tours*, for example, before the 1790s. Focusing on that sense of variety brought out by the *British Critic*, this section outlines the range of sources in which references to the *Tours* could be found by the turn of the century – a range just as diverse as some readers found the tours themselves.

In an early review that appeared in 1772, *The British Magazine and General Review* praised the 'industry and ingenuity' of the first Scottish *Tour* – terms that are often found in reactions to Pennant's work.[4] He had, the review ran, represented Scotland 'in a free, new, easy, and masterly manner', and it seems clear from this piece that Pennant's wide-ranging interests and loose, digressive method of recording his travels were part of the pleasure of reading his work. Yet, modern readers may (especially at first) find difficulties in the *Tours'* free and easy structure and miscellaneous content.[5] Mary-Ann Constantine has recently summed up the experience of reading Pennant's travel writing as one of being progressively weighed down with information ('historical, genealogical, economic, and scientific') to the point that 'it seems neither he nor his readers are moving forward at all'.[6] But to judge from reviews, Pennant seemed less meandering or Shandyesque to his contemporaries than he does to a modern audience accustomed to read for the plot. Miscellany was, after all, a peculiarly eighteenth-century format, and arguably the predominant model of cultural collection for the second half of the eighteenth century.[7]

In terms of reception history, the motley design of the *Tours* lends itself to wide patterns of reading, quotation and reprinting. Passages from the Welsh and Scottish *Tours* could, for example, be pulled from their original contexts and reproduced as stand-alone pieces. A section on remedies for minor ailments from the second Scottish *Tour*, was printed in the *Farmer's Magazine, and Useful Family Companion* in 1776, on the grounds that 'the subject matter is exceeding useful, and the book but in few Farmers' libraries'.[8] The blend of entertainment and instruction highlighted by the *British Critic* can be seen again and again in references to the tours from descriptions of place and scenery, to contemporary scientific works, and tracts on political economy: in 1793, *The Landscape Magazine* reprinted Pennant's description of Snowdon; Richard Watson, bishop of Llandaff, used Pennant in his five-volume *Chemical Essays* (1781–87, into a seventh edition by 1800), in his discussions of extracting silver from lead; Henry Gray Macnab, Professor of Moral Philosophy at Glasgow University, cited him in agricultural discussions about practices of burning turf and peat in the Hebrides in *Letters Addressed to the Right Honourable William Pitt [...] pointing out the inequality, oppression, and impolicy of the taxes on coal* (1793).

References such as these suggest another way of understanding miscellany in the period, as information unbounded by disciplinary categories. Different subjects overlap in a broad canvas of knowledge in Pennant's writing, in an age before disciplinary divisions route aesthetic, scientific, economic and political thought along separate lines. As William Mavor's turn to 'rays of information' and 'blended lights' suggests, Enlightenment ideas lie behind this conception of knowledge, and behind miscellaneous collecting as a means to improvement and progress. As Alexander Murdoch has shown, the eighteenth century witnessed a growing recognition of 'the pursuit of knowledge' as a 'civic virtue', a process accelerated by the spread of print culture, and largely concentrated in 'a commercial, urban, middle-class culture of paternalistic Whig elites'.[9] The circulation of print also had significant consequences for provincial Enlightenments, including 'the establishment of more localized cultures, as in the revival of Welsh, the resurgence of Scots song and poetry, and the emergence of regionally distinct English poets and working-class artists'.[10] I discuss some of the effects of the Enlightenment in Wales in the final section of this chapter, but saying more about the relations between amassing and organizing knowledge and place or region will help to anchor that discussion in wider eighteenth-century contexts.

Although the Welsh and Scottish *Tours* foreground place, they emerge from and are linked to concepts and movements beyond those places. For Stephen Copley, Pennant's travel writing should be grouped with the 'utilitarian descriptions' of a text such as Arthur Young's *A Six Months Tour through the North of England* (4 vols., 1770), and, more broadly, with the period's emergent forms of agricultural surveying and urban and county history. Copley explains that writers of regional surveys and other localized accounts were intent on bringing out a particular region's 'natural features, history, social structures, agricultural and manufacturing economies, and trades', and the conjectured effects of all these elements on 'the manners of the inhabitants'.[11] While generic differences distinguish texts structured as itineraries from county-based reports published by the Board of Agriculture, in the Enlightenment context of knowledge exchange, all things could be, and were, brought into the project of reading and recording eighteenth-century society. It is as part of this wider discourse that Pennant's *Tour in Wales* was a record of 'the changes then beginning to affect the culture and traditions of Welsh society', as R. Paul Evans has noted.[12] At the same time, however, separations of method and approach were visible in the period, even though the miscellaneous or blended form is so widely used and understood, as well as being the defining feature of Pennant's tours and their reception. For example, the accumulated detail of forms such as county and urban history marked them out from the conjectural histories being written by major figures within the Scottish Enlightenment. As

Rosemary Sweet has pointed out, this latter version of history was long the dominant one associated with the second half of the eighteenth century – a period in which history writing first began to be professionalized – to the relative neglect of accounts of places and the material evidence of their pasts constructed by antiquarians.[13]

In an important account, Sweet argues that eighteenth-century antiquarianism should be seen as 'the essential empirical foundation for the science of society [...] providing the factual ballast to the conjectural vessel of historical narrative'.[14] Reconstructing history through material detail had ideological and practical dimensions, in which modern citizens could improve their own society by studying the systems and beliefs of past societies. In this context, William Mavor's representation of travel writing as revelatory and improving conveys the way in which eighteenth-century antiquarianism was perceived as a theatre of public usefulness. Writing to Pennant in 1782, the antiquarian Richard Gough described the spirit of his work as that of popular Enlightenment: 'I wish to circulate knowledge as far as lies in my power',[15] and as Sweet further explains, 'the themes of patriotism and public service recur throughout his works'.[16] The point also applies to Pennant, but when the *British Critic* praised him for writing with 'the true spirit of an antiquarian',[17] in his lingering descriptions of Canterbury, Dover, Hastings and other historically freighted points in the *Journey from London to the Isle of Wight*, it was skimming the surface of a complex and contested set of views about the role and methodologies of history writing.

Both travel writing and antiquarian writing held the capacity for mapping the nation and determining its current state of progress; as in Pennant's work, sometimes in the same text. The reception of the Scottish *Tours* includes antiquarian-centred responses such as Charles Cordiner's *Antiquities and Scenery of the North of Scotland, in a series of Letters to Thomas Pennant, Esqr.* (1780), which extended Pennant's tour to the Highlands, and *Remarkable Ruins, and Romantic Prospects, of North Britain* (1788). In his preface to *Scotland Delineated, or a Geographical Description of Every Shire in Scotland* (1791), Robert Heron records his debt to Pennant and William Gilpin, 'ingenious writers' whose accounts of 'the grand or beautiful natural objects, the fine remains of antiquity, and the specimens of modern elegance, that embellish this country' had informed his work.[18] Other responses to the Scottish *Tours* took Pennant's portrait of contemporary Scotland as their subject. In 1775 *The Caledoniad*, a multi-volume collection of poems by Scottish authors, published 'A DESCRIPTION of the HUTS in ILAY from Mr. PENNANT's Voyage to the HEBRIDES, in the Year 1772', a 16-page reworking of the second *Tour's* account of Islay. The poem is a summary of Pennant's depiction of extreme poverty in Scotland, which plays down a topographical and geographical sense of place (found mainly in references in the footnotes) in order to emphasize the 'scenes of wretchedness', and 'direful Penury, Disease and Grief', in the original.[19] This rewriting of the *Tour* to the Hebrides particularly stands out as a response to Moses Griffith's engravings, as the poem was clearly written as much from the images – what the author terms the 'sculptur'd page'[20] – as from the text. The *British Magazine*'s 1772 review of the first Scottish *Tour* mentioned earlier, had commented on the visual depiction of landscapes, castles, ruins, and animals in Griffith's drawings, so striking because it was the first fully illustrated Scottish tour to be published. The *Caledoniad* poem, steeped in the illustrations ('A groupe which Pennant's mimic pencil draw […] now adorns his page, attracts our view'), offers some proof that the visual element of the tours made at least as much impact as the written one.[21]

Not by Victory Crowned: History and Tourism in 1790s Wales

Pennant's and Johnson's *Tours* defined Scotland for their own time, and for the tourists who followed them. Though they were not writing conjectural accounts, Penny Fielding has argued that the historical and anthropological approaches of their tours 'produced a diachronic Scotland whose land- and townscapes inscribed a history of people moving uncertainly in their stadial progress towards advanced agricultural practices'.[22] That halting sense of progress and uncertainty about the present can be felt in the *Caledoniad* poem, but the case is different for Wales, where uncertainty from the past becomes an important theme for writers in the 1790s who view Welsh landscape and history through the lens of Pennant's tours.

One of the effects of the study of antiquity in the eighteenth century was to make Britain's interlocking histories easier to see: for antiquarians like Richard Gough, it was the means to imagining a continuous British history.[23] This is a key point in the context of Wales, and the way in which Pennant and his reception history contributed to the revival of national sentiment. In their use and representation of the past, antiquarian and travel writers are again on common ground. As Benjamin Colbert has recently argued, eighteenth-century readers 'look to home tourism and travel writing more widely to […] connect the past and present in a single narrative.[24] Pennant himself was quickly written into that narrative by contemporary antiquarians, who drew on his work for local detail and descriptions of places and buildings, including monasteries, castles, and churches. The Scottish *Tours* reappear, for example, in Joseph Cooper Walker's *Historical Memoirs of the Irish Bards* (1786), while Edward Jones's *Musical and Poetical Relicks of the Welsh Bards* (1784) is a notably early example of the response to the 1778 *Tour in Wales*.

The period after 1800 sees a sense that an increasingly canonical Pennant is part of the fabric of the Welsh past; becoming historical himself in a new sense. In the 1790s, however, a decade marked by the pressures of wartime and widespread political uncertainties, the Welsh tours resonated more uneasily with tourists in Wales. The home tour movement was, so the classic account runs, boosted by war with France, which closed the Continent to would-be Grand Tourists.[25] New work on the wider eighteenth-century contexts of home tourism, and its development both from within Britain and under the influence of global travel, especially Pacific exploration, may diminish the persuasiveness of this narrative, but relations between Britain and France in general in the 1790s remain one of the most powerful, if often latent, frameworks for the domestic tour. As Colbert points out, tours of Wales, Scotland, and Ireland (at least after 1801) were constantly unsettled by the split or doubled nature of the domestic tour, which took place both on home ground and on territory that can seem linguistically, culturally, and topographically foreign. As a result, the home tour offers alternating states of alienation and assimilation, following 'localized itineraries that indicate a desire to discover closer at hand what is unfamiliar, yet at the same time to harmonize, homogenize, and extend the purview of home'.[26]

The discovery of the unfamiliar, and with it the 'indelibly inscribed' boundaries of nation within the British Isles,[27] leaves its mark on poetry written in response to Pennant. As the example from the *Caledoniad* suggests, poets drew on the *Tours* from virtually their first publication. Numerous references to the Scottish *Tour*, placing Pennant as an authority on natural history and Scottish culture, appear in Thomas Gisborne's *Walk in a Forest* (1796) and Thomas Maurice's *Poems, Epistolary, Lyric, and Elegiac* (1800). His descriptions of places and monuments are recycled in John Pinkerton's edition of John Barbour's *The Bruce; or the history of Robert I. King of Scotland. Written in Scottish Verse* (1790), and later Pennant would appear in Walter Scott. But the sense in which quotation is never a neutral practice is especially important in 1790s responses to the Welsh *Tours*, in which Pennant's work provides scaffolding to, permission for, and sometimes challenges to, other writers' anxieties and aspirations. The following section takes up two such responses.

In 1794, the poet and translator William Sotheby first published the long poem 'Llangollen' as part of an expanded version of *A Tour through parts of Wales* (1790). Dated to 'the close of the autumn' of 1792, this 16-page poem in heroic couplets takes the form of a journey through Llangollen in northeast Wales. Two years later the Lichfield poet who had risen to prominence in the 1780s, Anna Seward, published 'Llangollen Vale' (1796), the fruit of her tour across north Wales in the summer and autumn of 1795. Both of these long poems are set in and around Llangollen and use Pennant's version of the region's past, which turns on the medieval wars fought, and ultimately lost, for Welsh independence. The central figure in this history is Owain Glyndŵr, whose place in Pennant's *Tour in Wales* is discussed elsewhere in this volume by Dafydd Johnston. Glyndŵr appears around halfway through Sotheby's elegiac and autumnal poem, which is an exercise in contemplating the passage of time and the sorrowing self, as in this description of the ruins of Valle Crucis abbey, just outside Llangollen:

> Sorrowing I turn, and through the birchen shades,
> That sweep o'er Llandysilio's shelter'd glades,
> Seek the deserted fane [abbey] [...]
> Time's mellowing damps that moulder where they fall,
> Stream in rich stains, and picture o'er the wall.
> I pause – to voluntary woes resigned,
> And lenient grief that leaves a balm behind.[28]

The elegiac mood extends to Sotheby's treatment of Glyndŵr, imagined in this poem as worn down by a war that he cannot possibly win: 'I trace the caves, and deep recesses hoar / That roofed the war-worn head of wild Glendore' (114).

Pennant's account of Glyndŵr underpins the opening passages of Seward's 'Llangollen Vale', the first half of which recounts the history of Llangollen as a bloody struggle for Welsh freedom. The second half of the poem is a tribute to Lady Eleanor Butler and Sarah Ponsonby, the Ladies of Llangollen, living in famed retirement, quiet industry, and what Seward terms 'sacred FRIENDSHIP'.[29] It is difficult to reconcile the two halves of the poem, though the shift from ideas of war to those of peace, science and taste (7) puts the refining influence of women at the heart of the poem: 'What boasts Tradition, what th' historic Theme [...] in this Cambrian Valley, Virtue shows / Where, in her own soft sex, its steadiest lustre glows' (9). The war scenes borrowed from Pennant are, however, not easily dismissed. The year of Seward's Welsh tour, 1795, was a year of corn and anti-press gang riots in Wales, a period of famine and protest, while the French wars continued (with a break in 1801–3) to 1815.[30] In this context, Seward's Llangollen appears a fantasy of settled order that denies divisions of place, class and race.

By the 1790s Pennant was a public conservative, closely involved in local movements to resist French Revolution–inspired disorder in Wales.[31] Sotheby and Seward are not radical figures either, though Seward's letters from this period reveal her private unease about the war with France, which she could not have had made public without appearing critical of the Tory government's conduct in the 1790s.[32] Pennant's exact place in the overlapping views of past and present, polemical refinement (Seward) and contemporary uncertainty (Sotheby), that characterize 1790s views of Llangollen is not easy to pin down. However, Mary-Ann Constantine has recently described his contribution to the topographical tradition in his Welsh *Tours* as one in which landscape 'becomes the occasion for historical suggestion, a continual revelation of sites of contest and controversy'.[33] This point gets to the heart of the problem: Pennant's *Tours* both enable those who follow him through north-east Wales to explore their doubts about what was happening in the 1790s, and compound those doubts with historical examples of failure – the Welsh quest for freedom is 'not by victory crowned' (114), Sotheby reflects – and British disunity. 'Llangollen' vividly captures what it may have felt like to live through an unstable succession of events, as in the early 1790s, although Sotheby's anxieties about the present are also always in the process of being displaced into history. His style in the poem reflects this sense of deferral or deflection, in fluid topographical writing which tries (but is always conscious of failing) to capture the overwhelming breadth of the view, the swirling River Dee, and the constant shifts of colour, light and shadow across the landscape.

Sotheby's emotionally misted-up poem speaks to the fears and occlusions of the 1790s that unsettle English home tourists, but the history of north Wales is put to very different uses in works produced from within Wales. Rosemary Sweet has argued that '[a] sense of the past and historic identities were essential features in the imagined communities of eighteenth-century nationalism'.[34] This argument will also gloss early nineteenth-century responses to Pennant, which draw on the patriotic sentiments of his Welsh *Tours* – as, for example, in his representation of Owain Glyndŵr as a national hero – in new works written from cultural nationalist perspectives. However, as Dafydd Johnston shows, the effect of Glyndŵr in the *Tours* is less political than 'concerned with the honour of the Welsh people'. From this perspective, late nineteenth- and twentieth-century perceptions of Glyndŵr as the embodiment of a native tradition of resistance can only be seen as an ironic misinterpretation of Pennant's project.[35] That said, the *Tours* were of course put to new uses virtually from their first appearance, and the final section of this essay sets them in their early nineteenth-century contexts.

Future Minstrelsy: Rethinking Bardic Nationalism with Pennant

The turn of the nineteenth century saw a trend for collections of travel writing: in addition to anthologies like Mavor's *British Tourists*, major multivolume topographical works appeared in this period, including Edward Wedlake Brayley and John Britton's *The Beauties of England and Wales* (18 vols, 1801–18), and Richard Ayton and William Daniell's *A Voyage Round Great Britain* (8 vols, 1814–25). Benjamin Colbert describes these as 'encyclopedic projects' full of antiquarian detail that 'stand for and promote a textualized landscape subordinate to a grand narrative'.[36] In these accounts, place, locality, history and tradition bind the edges of a totalizing vision of Britain. And yet, Colbert's summary suggests that it is possible to see new perceptions of nation emerging from the closely detailed 'textualized landscape' of the *Tours* from behind the grand narrative of union that collections and circumnavigations of the British Isles otherwise seem to be staging or confirming. The final section of this chapter expands this point through early nineteenth-century uses of Pennant's Welsh *Tours*, suggesting that quotation, often within the scholarly apparatus of annotation, contributes to a cultural nationalism in Wales in which future development appears increasingly urgent.

Weighty, often beautifully illustrated, nineteenth-century sets of *Tours* (Daniell's is a good example) pose the idea that the home tour 'is not an act of superficial dilettantism, but amounts to a reading of the nation and the foundations on which it is built'.[37] Paul Smethurst has recently set Pennant's *Tours* in a trajectory of Unionist travel texts, which, at least from Defoe in the 1720s to the end of the century, project the united condition of the British Isles.[38] And yet the reception of the Welsh *Tours* after 1800 includes a patriotic revisionist strand that imagines a revived future Welsh nation, as well as summing up what that nation is and was. In the 1790s the state of the nation looked dark, unclear at best, to William Sotheby; but the horizon is brighter for Welsh writers in the period, even though they concede there is work to do.

The response to Johnson, which Katie Trumpener reads as 'a moment of national trauma, which made visible [...] the long-term impossibility of full Scottish integration into Britain', shows that the home tour was always potentially a conflicted space in this period, a space for airing claims and counter-claims.[39] Pennant's work, too, spans contested ground. Even though he was part of the Anglicized gentry, and though his work has been seen as part of a longer attempt to hold the union steady through periods of revolution and war, the *Tour in Wales* can sound provocatively patriotic. The grandstanding of its opening is a good example of this element of the *Tour*. 'I now speak of my native country, celebrated in our earliest history for its valour and tenacity'.[40] John Barrell has recently suggested that this style of address reflects 'a new confidence about Wales and the culture of Wales, addressed to a wider British audience without apology or deference'.[41] Class clearly makes a difference in this pose. In his gentry status, Pennant was untypical of the men of more modest backgrounds who largely dominated the artistic life of late eighteenth-century Wales, but it is likely that he was able to appear before the public with such confidence because of his class as well as his reputation. The same tone did not come easily, for example, to the Welsh painter and travel writer Edward Pugh, son of a market-town barber, who 'probably had [Pennant's *Tours*] open on his desk' as he wrote up his own tour *Cambria Depicta* (1815).[42]

And yet, the *Tour*'s patriotic gestures offer later readers challenges and opportunities. Sparky moments such as the opening belong to the wider cultural revival of late eighteenth-century Wales.[43] Secession from the union rarely if ever features in that movement, but the specificity of the Welsh revival is an important reminder that national and regional difference never disappeared in this period. Even though union was 'the dominant political tendency' of the century, as Martin Fitzpatrick points out, 'the Enlightenment in Britain was never unified in this way'.[44] Even the Scottish Enlightenment may be seen as the means by which the Scots both defend their particularity and difference, and claim a distinct place within Britain: 'no less than a project by leaders of the Scottish professional middle classes to reinvent their ancient country as a dynamic part of a modern and expanding British polity'.[45] As Vladimir Jankovic has argued, provincial Enlightenment was a means of actively resisting, not necessarily imitating or even adapting, metropolitan practices. For Jankovic, the emergence of 'specifically *local* scholarship', including Pennant's, on subjects from natural history to geology and regional weather in this period should be seen in the context of, and continuous with, earlier chorographic concepts of place and identity, rather than as 'uncritical *acceptance* of [...] urban mores'.[46]

Wales had its own version of Enlightenment, shaped by religion, print culture, attitudes towards the national past, and calls for progress and improvement influenced by localized issues of economy and social justice.[47] Its Enlightenment was more radical and artisan-centred than in Scotland, but Pennant's *Tours* also raise the profile of a distinct and historicized Wales within the union – a point that the tagline sometimes attached to his name, 'the father of Cambrian tourists', suggests. References to the Welsh *Tours* suggest that Pennant is a fixed mark in the cultural landscape by at least the 1810s. With a little distance, the *Tour in Wales* becomes visible as a patriotic watershed or waymarker, a foundation on which to imagine a future Welsh nation as well as a scholarly source for later poets and travel writers. Footnotes provide the main route by which later writers negotiate Pennant's text, and from the *Caledoniad* to Anna Seward, annotation is the main channel through which his observations and ideas move through time. Looking at uses of Pennant in footnotes beyond 1800, then, I conclude this chapter with responses to the Welsh *Tours* in poems and songs by Charlotte Wardle and Felicia Hemans.

Charlotte Wardle was born near Mold in north-east Wales, daughter of the MP Gwillim Lloyd Wardle, who was fêted as a public-spirited liberal when he successfully brought corruption charges against the Duke of York in 1809 (though his success lasted only until his own political career was also brought down by corruption charges). Little is known of Charlotte Wardle, though her death in 1828 shortly after the birth of her first child suggests that she was probably born in the 1790s.[48] In 1814 she published a ballad romance in five cantos, *St. Aelian's, or the Cursing Well*. Set in north Wales, and based on a real cursing well near Conwy,[49] the poem is a medievalist tale of dynastic rivalry, jealousy, thwarted love and revenge. Its title refers to events in Canto IV where the villain of the piece uses a witch to curse his rival using a cursing well, which involves the desired victim's name being written on paper or slate and cast into the well. The curse can only be lifted when the name is retrieved from the well. Beyond the real-life referents of the poem, *St. Aelian's* is clearly in dialogue with contemporary writings, notably Walter Scott, and Robert Southey, whose 1802 supernatural ballad 'The Inchcape Rock' appears in endnotes to the poem.[50] As well as drawing on contemporary Gothic and medievalist writing, Wardle looks to Pennant's *Tours*, which locate the poem in north Wales via authoritative descriptions of buildings, such as Conwy castle, natural scenery, such as the Nant Ffrancon pass in Snowdonia, and not least St Aelian's well itself (a quotation from Pennant explains that '[i]t was resorted to by the Welsh to call imprecations and the vengeance of the Saint on any one who had done them an injury').[51] Compared with the supernatural machinery of the main text, the references to Pennant also locate the poem within a matrix of scholarly exchange that prepares the ground for the poem's serious concluding remarks (discussed below).

My final example is the Liverpool-born poet Felicia Hemans (1793–1835), who moved to north Wales at the age of seven, and whose writing career was shaped by her lived experience of Wales, as I have argued elsewhere.[52] We can see her directly responding to Pennant in the texts of her little-studied 1822 collaboration with the musician John Parry, *A Selection of Welsh Melodies*. As in poems by Sotheby, Seward and Wardle, annotation builds up a patchwork of antiquarian references, though Hemans ranges more widely than her predecessors, basing her songs for *Welsh Melodies* on Edward Jones's *Musical and Poetical Relicks of the Welsh Bards*, the ninth-century song cycle, *The Heroic Elegies of Llywarç Hen* (1792), as translated by the editor and lexicographer William Owen Pughe, *The Cambrian Biography* (1803 – also edited by Owen Pughe), and Edward Davies's *Celtic Researches on the Origin, Traditions and Languages of the Ancient Britons* (1804), as well as Pennant's Welsh *Tour*. Pennant is, however, the single strongest influence in this text. For example, the tenth song in *Welsh Melodies*, 'Owen Glyndwr's War Song', is – like the opening passage of 'Llangollen Vale' – pure Pennant, and footnoted as such:

Saw ye the blazing star*?
The heavens look down on freedom's war,
    And light her torch on high!
Bright on the dragon crest†
It tells that glory's wing shall rest
    When warriors meet to die!
Let earth's pale tyrants read despair
    And vengeance in its flame.
Hail ye, my bards! the omen fair
    Of conquest and of fame,
And swell the rushing mountain-air
    With songs to Glyndwr's name.[53]

* The year 1402 was ushered in with a comet or blazing star, which the bards interpreted as an omen favourable to the cause of Glyndwr. It served to infuse spirit into the minds of a superstitious people; the first success of their chieftain confirmed this belief, and gave new vigour to their actions. – *Vide* PENNANT.[54]

† Owain *Glyndwr* styled himself the *Dragon*, a name he assumed in imitation of *Uther*, whose victories over the Saxons were foretold by the appearances of a star with a dragon beneath, which Uther used as his badge; and, on that account it became a favourable one with the Welsh. – PENNANT.[55]

As in Wardle's poem, the stylistic spilt between text and footnote in Hemans's songs is revealing. The emotion and melodrama that at times characterizes the song texts are countered by or grounded in the scholarly conversation of the footnotes. That conversation is centrally indebted to William Owen Pughe, a figure at the heart of the Welsh cultural revival, who dedicated his translation of *Llywarç Hen* to Thomas Pennant, and who brought Hemans and Parry together to work on their national song project. The scholarly scaffolding that lies behind the songs' depiction of the Welsh national past transforms their meanings, and especially their status as comments on contemporary Wales. A clearly stated desire to revive the Welsh nation links Charlotte Wardle in 1814 and Felicia Hemans in 1822 (in the case of Hemans this can also be seen in poems dating to 1810–12).[56] At the close of *St. Aelian's*, Wardle presents herself as a champion of Welsh culture, writing in character as a female minstrel who aspires to bring in a modern age of liberty and national consciousness:

Thy [Cambria's] sons are bold, and still retain
Remembrance of the sanguine plain,
That oft their forefathers have strew'd
With hills of slain and hostile blood;

And still their feats are proudly sung,
In numbers of their native tongue.
Oh! may their sons recorded be,
In strains of future minstrelsy [...] [96]

Perhaps surprisingly, there is no British dimension to this 60-line concluding address: the nation she encourages and exalts in the poem's final line ('My native country! fare thee well!') is specifically Wales.

Wardle's confidence to speak out in this passage, coming at the end of a poem notable for its combination of the supernatural and the scholarly, suggests that she is setting patriotism and memory in new relation or formations. But how should we interpret her vision of 'future minstrelsy'? Can we read this conclusion as a kind of bardic nationalism? On one level, both Wardle and Hemans sound like national bards, turning to displays of deep emotion and stock strategic symbols of nation, such as harps, that are key elements of bardic nationalism. Missing here, however, is that mixed sense of mourning, nostalgia, and defensiveness – 'sublime, heroic, and tragically doomed' – that Katie Trumpener has identified as central to bardic nationalism.[57] Discussing James Macpherson's *Ossian* poems of the 1760s, Trumpener focuses on the barrenness of the present in those texts, smothered by the history of loss that largely defines the mood of bardic nationalism: 'Left in a present that is empty, [Ossian] revives the past to fill it.'[58] More recent work on the specifically Welsh elements of Celtic-peripheries history has, however, shown that a model of bardic history based on nostalgia and deficit does not account for the situation in Wales.[59] Pennant himself is a case in point; his depiction of the bards as figures who 'animated' and sustained the Welsh in their long struggle for liberty draws on a wider understanding of the Welsh bard as guardian of the historical, moral and mythological knowledge that acts as a foundation for his society or community and its future development.[60]

There are points when calls for a renovated, rejuvenated Cambria sit alongside the melancholy and elegiac, in Hemans's song texts in particular, but the evidence of poetry inspired by Pennant between 1790 and 1822 suggests a split in the responses of writers who were visiting Wales, and those who were Welsh (or Welsh by adoption). For Sotheby and Seward, Pennant represents what Mary-Ann Constantine calls 'dangerous history', residual or imagined in the landscape as a projection of the conflict and turbulence of the 1790s.[61] However, Trumpener's model of loss through history, leading to an empty present, is not what Pennant, with his high tide of antiquarian detail and Enlightenment eye to the spread of knowledge and advancement of society, represents for writers with stronger Welsh connections than William Sotheby or Anna Seward. Felicia Hemans's footnotes, rather than her necessarily theatrical main texts – which are songs to be performed after all – discard the narratives of emotion and sentiment that characterize *Ossian* in favour of an emergent tradition of authoritative, learned detail. And the differences between early reviews of Pennant, responses by travellers to Wales in the 1790s, and Welsh poetry by women after 1810, demonstrate how his reception changes over time. Drawing on the work of Colin Kidd, Rosemary Sweet has shown how antiquarians developed the history of nationalism in the period via their discovery of the plural background of British identities: 'acutely conscious of origins', antiquarians 'did not subscribe to a notion of homogenous nationhood'.[62] Pennant's project is part of the movement outlined by Sweet, but the consequences for Wales can be pushed further because revisionist or dissenting accounts emerge from the heterogeneous detail that defines antiquarian works, even when they are also contributing to grand narratives such as the British union. In this way, the antiquarian method, sometimes seen as pedantic or indiscriminate in its approach to knowledge, holds the key to alternative narratives. As detail fragments into local perspectives on history, place and people it may tend towards an unmanageable particularity, but so it also provides a foundation on which to create a scholarly discourse and enlightened nationalism for nineteenth-century Wales.

Notes

1 William Mavor, *The British Tourists; or Traveller's Pocket Companion, through England, Wales, Scotland, and Ireland*, 6 vols (London, 1798), I: v. Further references to this volume are included in the text.

2 Paul Smethurst, *Travel Writing and the Natural World, 1768–1840* (Palgrave, 2012), 112.

3 *The British Critic, for July, August, September, October, November, and December. MDCCCI* (London: F. and C. Rivington, 1801), 580.

4 *The British Magazine and General Review [...] Volume I* (London, 1772), 60.

5 Elizabeth Zold focuses on the difficulties encountered by new students of eighteenth-century travel writing, and suggests that their resistance to these texts' apparent lack of narrative order is a common problem: 'Discomforting Narratives: Teaching Eighteenth-Century Women's Travelogues', *ABO: Interactive Journal for Women in the Arts, 1640–1830* 4, issue 2, http://scholarcommons.usf.edu/abo/vol4/iss2/3 (accessed 4 November 2014).

6 Mary-Ann Constantine, '"To trace thy country's glories to their source": Dangerous History in Thomas Pennant's *Tour in Wales*', in *Rethinking British Romantic History, 1770–1845*, ed. Porscha Fermanis and John Regan (Oxford: Oxford University Press, 2014), 121–43. I am grateful to the author for allowing me to read this chapter prior to publication.

7 A point recently emphasized by the publication of the online *Digital Miscellanies Index*; see http://digitalmiscellaniesindex.org/ (accessed 4 November 2014).

8 *The Farmer's Magazine, and Useful Family Companion*, 5 vols (London and Bath, 1776), I: 105–7. See Thomas Pennant, *A Tour in Scotland, and Voyage to the Hebrides, MDCCLXXII*, 2 vols (London, 1776), II: 41–42, for the original passage.

9 Alexander Murdoch, 'A Crucible for Change: Enlightenment in Britain', in *The Enlightenment World*, ed. Martin Fitzpatrick et al. (London and New York: Routledge, 2007), 104–16 (113, 110).

10 Ibid. 110.

11 Stephen Copley, 'William Gilpin and the Black-Lead Mine', in *The Politics of the Picturesque: Literature, Landscape, and Aesthetics Since 1770*, ed. Stephen Copley and Peter Garside (Cambridge: Cambridge University Press, 1994), 42–62 (42).

12 R. Paul Evans, 'Thomas Pennant (1726–1798): "The Father of Cambrian Tourists"', *Welsh History Review* 13, no. 4 (1987): 395–417 (412).

13 Rosemary Sweet, 'Antiquaries and Antiquities in Eighteenth-Century England', *Eighteenth-Century Studies* 34, no. 2 (2001): 181–206.

14 Ibid. 188.

15 Quoted in ibid. 189.

16 Ibid. 186.

17 *British Critic*, 583.

18 Robert Heron, *Scotland Delineated, or a geographical description of every shire in Scotland* (Edinburgh and London, 1791), 5.

19 'A Clergyman', 'A Description of the Huts in Ilay from Mr. Pennant's Voyage to the Hebrides, in the Year 1772', *The Caledoniad: A Collection of Poems, Written Chiefly by Scottish Authors*, 3 vols (London, 1775), III: 188.

20 Ibid. 196.

21 Ibid. 191. A footnote to these lines directs readers to the source: 'See the plate, p. 229, exhibiting the inside of a weaver's cottage in Ilay.'

22 Penny Fielding, 'Burns's Topographies', in *Scotland and the Borders of Romanticism*, ed. Leith Davis, Ian Duncan and Janet Sorenson (Cambridge: Cambridge University Press, 2004), 170–87 (170).

23 Sweet, 'Antiquaries and Antiquities': 187.

24 Benjamin Colbert, 'Introduction: Home Tourism', in *Travel Writing and Tourism in Britain and Ireland*, ed. Benjamin Colbert (London: Palgrave Macmillan, 2012), 1–12 (4).

25 See, e.g., Esther Moir, *The Discovery of Britain: The English Tourists, 1540–1840* (London: Routledge and Kegan Paul, 1964); Malcolm Andrews, *The Search for the Picturesque: Landscape Aesthetics and Tourism in Britain, 1760–1800* (Aldershot: Scolar Press, 1989).

26 Colbert, 'Introduction', 1.

27 Ibid. 1–2.

28 William Sotheby, 'Llangollen', *A Tour through parts of Wales, Sonnets, Odes, and other Poems. With Engravings from Drawings taken on the Spot, By J. Smith* (London, 1794), 116. Further references by page number are included in the text.

29 Anna Seward, 'Llangollen Vale', *Llangollen Vale, With Other Poems* (London, 1796), 6. Further references by page number are included in the text.

30 David J. V. Jones, *Before Rebecca: Popular Protests in Wales, 1793–1835* (Cardiff: University of Wales Press, 1973).

31 John Barrell, *Edward Pugh of Ruthin 1763–1813: 'A Native Artist'* (Cardiff: Cardiff University Press, 2013), 30.

32 Claudia Kairoff Thomas, *Anna Seward and the End of the Eighteenth Century* (Baltimore: Johns Hopkins University Press, 2012), 98–116, shows that, for Seward, to have gone public with her views would have risked representing her as a much more oppositional figure than she really was.

33 Constantine, '"To trace thy country's glories to their source"', 123.

34 Sweet, 'Antiquaries and Antiquities': 181.

35 See Dafydd Johnston's contribution to this volume, 105–21 (114).

36 Colbert, 'Introduction', 4.

37 Ibid.

38 Paul Smethurst, 'Peripheral Vision: Landscape, and Nation-Building in Thomas Pennant's Tours of Scotland, 1769–72', in *Travel Writing and Tourism*, ed. Colbert, 13–30.

39 Katie Trumpener, *Bardic Nationalism: The Romantic Novel and the British Empire* (Princeton, NJ: Princeton University Press, 1997), 33.

40 Thomas Pennant, *A Tour in Wales* (London, 1778), 1.

41 Barrell, *Edward Pugh*, 2.

42 Ibid. 174.

43 Prys Morgan, 'From Death to a View: The Hunt for the Welsh Past in the Romantic Period', in *The Invention of Tradition*, ed. Eric Hobsbawm and Terence Ranger (Cambridge: Cambridge University Press, 1983), 43–100.

44 Martin Fitzpatrick, 'Enlightenment', in *An Oxford Companion to the Romantic Age: British Culture 1776–1832*, ed. Iain McCalman (Oxford: Oxford University Press, 1999), 299–310 (299).

45 Murdoch, 'A Crucible for Change', 112.

46 Vladimir Jankovic, 'The Place of Nature and the Nature of Place: The Chorographic Challenge to the History of British Provincial Science', *History of Science* 38, no. 1 (2000): 79–113 (80, 83). Original emphases.

47 For an illustration of this last point, see the introduction to Marion Löffler (with Bethan Jenkins), *Political Pamphlets and Sermons from Wales 1790–1806* (Cardiff: University of Wales Press, 2014).

48 She was probably born at some point after 1792, when Gwillim Lloyd Wardle married Ellen (or Elen) Elizabeth Parry.

49 Richard Suggett, *A History of Witchcraft and Magic in Wales* (Stroud: History Press, 2008), 119–22.

50 Charlotte Wardle, *St. Aelian's Well: A Tale: In Five Cantos* (London, 1814), 86. Further references to this volume are included in the text.

51 Ibid. 105, 107, 110.

52 Elizabeth Edwards, '"Lonely and voiceless your halls must remain": Romantic-era National Song and Felicia Hemans's *Welsh Melodies* (1822)', *Journal for Eighteenth-Century Studies* 38, no. 1 (2015): 83–97.

53 [Felicia Hemans], 'Owen Glyndwr's War Song', *A Selection of Welsh Melodies, with Symphonies and Accompaniments by John Parry, and Characteristic Words by Mrs. Hemans* (London: J. Power, 1822), 38.

54 See Pennant, *A Tour in Wales*, 321.

55 Ibid. 335.

56 See Edwards, '"Lonely and voiceless your halls must remain"', for a discussion of manuscript poems written in 1810–12.

57 Trumpener, *Bardic Nationalism*, 76.

58 Ibid.

59 Sarah Prescott, *Eighteenth-Century Writing from Wales: Bards and Britons* (Cardiff: University of Wales Press, 2008), 57–83, outlines the antiquarian–poetic resistance in Wales to the bard as a sublime figure of loss and defeat. Mary-Ann Constantine discusses the stonemason–poet Edward Williams's (Iolo Morganwg) self-identification as proof of a surviving tradition, and as a resolutely oppositional and future-looking Welsh bard

in *The Truth Against the World: Iolo Morganwg and Romantic Forgery* (Cardiff: University of Wales Press, 2007), and, with Elizabeth Edwards, '"Bard of Liberty": Iolo Morganwg, Wales and Radical Song', in *United Islands? The Languages of Resistance*, ed. John Kirk, Andrew Noble and Michael Brown (London: Pickering and Chatto, 2012), 63–76.

60 Quoted in Johnston this volume, 116.

61 Constantine, '"To trace thy country's glories to their source"', 122–23.

62 Sweet, 'Antiquaries and Antiquities': 197.

Part II

NATURAL HISTORY AND THE ARTS

Chapter 8

'AS IF CREATED BY FUSION OF MATTER AFTER SOME INTENSE HEAT': PIONEERING GEOLOGICAL OBSERVATIONS IN THOMAS PENNANT'S *TOURS* OF SCOTLAND

Tom Furniss[*]

Thomas Pennant's *A Tour in Scotland, 1769* (1771) and *A Tour in Scotland, and Voyage to the Hebrides, MDCCLXXII* (1774, 1776) played a key role, along with James Macpherson's Ossian poems of the 1760s, in representing the Highlands and Islands of Scotland as attractive places for a growing number of travellers and tourists. Brian D. Osborne claims in his introduction to a modern edition of *A Tour in Scotland, 1769* that 'to Pennant must go much of the credit for the non-Scottish world's discovery of Scotland and its development as a destination for travellers in search of the picturesque'. For Osborne, Pennant's appreciation of 'the grandeur and natural beauty of Scotland' made him 'one of the first generation of travellers who saw in wild mountain scenes anything more than an awful wilderness'.[1] Yet, as Charles W. J. Withers stresses in his introduction to a recent edition of *A Tour in Scotland and Voyage to the Hebrides, 1772*, although 'Pennant was certainly aware of Scotland's visual grandeur, […] it is not a central theme of his *Tour*'.[2] Pennant's central theme, I suggest, is the natural history of Scotland, and his tours helped to make the Highlands and Islands into primary destinations for travellers in search of the natural history of the northern wilds of Britain. Pennant attended to all aspects of Scotland's natural history, but in this chapter I focus on his geological observations, especially his discovery of impressive basaltic formations in the Inner Hebrides and what he took to be the remains of ancient volcanoes in various parts of Scotland – discoveries for which Pennant has not received any credit. I will also suggest, however, that there are significant interconnections between Pennant's observations of Scotland's geomorphology and his aesthetic response to its landscape.

Pennant was one of the foremost natural historians in eighteenth-century Britain. According to G. R. de Beer, he was 'the leading British zoologist after [John] Ray and before [Charles] Darwin'.[3] Pennant's most important scientific work, *British Zoology*, published in London in 1766 and 1767 in four beautifully illustrated quarto volumes, 'was organized according to the classificatory systems of John Ray' and 'established him in the eyes of contemporaries as a leading European natural historian'.[4] Yet, despite Pennant's stature as a natural historian, little attention has been paid to his contribution to mineralogy and the early history of geology. Pennant is not mentioned at all in Rachel Laudan's *From Mineralogy to Geology: The Foundations of a Science* (1987) or in David Oldroyd's *Thinking about the Earth: A History of Ideas in Geology* (1996). Gordon L. Davies's *The Earth in Decay: A History of British Geomorphology, 1578–1878* (1969) mentions Pennant only once, identifying him as the first person to publicize the Parallel Roads of Glen Roy. Roy Porter's *The Making of Geology: Earth Science in Britain, 1660–1815* (1977) ascribes to Pennant only a minor role as a travel writer who was skilled in describing minerals, rock formations and landscapes. Martin J. S. Rudwick's *Bursting the Limits of Time: The Reconstruction of Geohistory in the Age of Revolution* (2005) sees Pennant merely as the conduit for Sir Joseph Banks's account of Staffa, while Dennis R. Dean's *James Hutton and the History of Geology* (1992) suggests that Hutton read *A Tour in Scotland, and Voyage to the Hebrides, MDCCLXXII*, but otherwise ignores Pennant.[5]

The defining concerns of eighteenth-century natural history – empirical observation, description and classification – were inherited from the Baconian natural history of seventeenth-century pioneers such as Ray and Francis Willughby.[6] Natural history defined itself against the cosmogonies or speculative theories of the earth that flourished at the end of the seventeenth century, focusing instead on the 'observation of Matter of Fact' and on exploring 'the most desert Rocks and Mountains, as [well as] the more frequented Valleys and Plains' in search of evidence and specimens.[7] Discoveries made in the field were supplemented by the reports of other travellers and through the use of circulated questionnaires. The goal was to produce natural histories of particular regions and to construct a general system of classification. Natural history divided up the objects of the natural world into animal, vegetable and mineral 'kingdoms'. Mineralogy set out to imitate the methods of observation, description and classification that had been so successful in the other kingdoms. Minerals, rocks and 'fossils' were treated as natural kinds that had existed in a relatively static state since the creation, and mineralogy did not therefore concern itself with theories about their formation or historical development. Mineralogy, then, was quite different from the geology that emerged in the late eighteenth century, which assumed that the earth was a dynamic system and studied the materials and forms of its crust in terms of their history and/or their causation.

Yet, although Porter distinguishes between mineralogy and geology in these ways, he also stresses the contribution that mineralogy made to the formation of geology.[8] Furthermore, the static view of the earth that characterized mineralogy began to be challenged in the second half of the eighteenth century by interpretations of strata, earthquakes and volcanoes that suggested that the earth had had an extremely long dynamic history.[9] The problematic goal of producing a Linnaean classification of rocks and minerals gave way to the attempt, by mineralogists such as Abraham Gottlob Werner, to classify them according to their supposed causes and temporal sequence of formation.[10] Mineralogy thus got caught up in contemporary controversies about the history and formative causes of the earth's crust.[11] While the Neptunist mineralogy developed by Werner and others interpreted the earth's materials and formations as largely resulting from the effects of one or more inundations by a universal ocean, Vulcanist mineralogy interpreted some of the same materials and formations as evidence for considerable volcanic activity over long periods of the earth's history.[12] Both sides in the dispute assumed that the goal of mineralogy was no longer static classification but the understanding of rocks and minerals in terms of their causation and historical sequence. Both Porter and Rachel Laudan therefore see mineralogy as anticipating, making possible, or even morphing into the geology that emerged at the turn of the century, a seismic paradigm shift in which '[t]he static had given way to the dynamic'.[13]

On the face of it, Pennant's responses to Scotland's geomorphology are fully governed by the paradigms of eighteenth-century natural history. In contrast to Romantic-period geological travellers of the following generation, who fashion themselves as quest heroes who overcome difficulty and danger in order to arrive at remote places and radical theories of the earth, Pennant is much more concerned with objectively observing and describing the particulars of Scotland's natural history, landscape, antiquities and customs.[14] Pennant's attention to geological features prioritizes observation and description over causal theory, and avoids attempting to understand their place within a history of the earth's evolving system of forces and materials. Pennant's admission in the first volume of *A Tour in Scotland, 1772* that the intricate folding of the strata of coal and stone at Whitehaven that he saw en route to Scotland on his second tour was due to 'Operations of nature past my skill to unfold' suggests, through a gentle pun, that theoretical speculation about the causal origins of geological features was beyond his capacity or outside his remit.[15] Whether or not Pennant is being overly modest here, the reluctance to engage in geotheory or geohistory was characteristic of the natural history of the period.

When Pennant visited Edinburgh during his first tour of Scotland, his observations of Arthur's Seat and Salisbury Crags – features that would soon become significant in the history of geology – oscillate between different modes of scientific analysis and aesthetic response:

> Near [Holyrood] palace are the *Parks* first inclosed by *James* V. within are the vast rocks known by the names of *Arthur's* Seat and *Salusbury's Craigs*; their fronts exhibit a romantic and wild scene of broken rocks and vast precipices, which from some points seem to over-hang the lower parts of the city. Great columns of stone, from forty to fifty feet in length, and about two feet in diameter, regularly pentagonal, or hexagonal, hang down the face of some of these rocks almost perpendicularly, or with a very slight dip, and form a strange appearance. Considerable quantities of stone from the quarries have been cut and sent to *London* for paving the streets, its great hardness rendering it excellent for that purpose. Beneath these hills are some of the most beautiful walks about *Edinburgh*, commanding a fine prospect over several parts of the country.[16]

When Pennant observes those 'great columns of stone', he employs the characteristic mode and terminology of mineralogical description. This careful description of the basaltic pillars of Salisbury Crags (an exposed sill) does not lead to speculation about their composition or origins. Similarly, there is no recognition that Arthur's Seat is part of an extinct volcano (it should be said that its volcanic provenance was not accepted by Edinburgh's geologists, including James Hutton, even after it had been pointed out by Faujas de Saint-Fond in 1784). Pennant's emphasis on the way the stone has been exploited for commercial purposes exemplifies the common link between mineralogy and economic exploitation. His use of the term 'fine prospect' to describe the view from the walks beneath Arthur's Seat and Salisbury Crags invokes the contemporary cult of the picturesque. But when he writes that the fronts of Arthur's Seat and Salisbury Crags 'exhibit a romantic and wild scene of broken rocks and vast precipices, which from some points seem to over-hang the lower parts of the city', he ventures into impressionistic hyperbole derived from the cult of the sublime. Yet, Pennant characteristically omits the sublime's defining characteristic – its impact on the perceiving subject.

The static view of nature assumed by both natural history and the picturesque were, however, tested to the limit by the mountains of the Scottish Highlands. There are moments in Pennant's tours when the sublime features of mountain topography seem inevitably to point to dynamic geological processes in the distant past. On his second tour, for example, Pennant allows himself some speculation about the forces that had shaped the astonishing mountain landscape of Ross-shire:

> Ascend a very high mountain [...] Pass under some great precipices of lime-stone, mixed with marble: from hence a most tremendous view of mountains of stupendous height and generally of conoid forms. I never saw a country that seemed to have been so torn and convulsed: the shock, whenever it hap-pened, shook off all that vegetates: among these aspiring heaps of barrenness, the sugar-loaf hill of *Suil-bhein* [Suilven] made a conspicuous figure: at their feet, the blackness of the moors by no means assisted to cheer our ideas.[17]

We might expect that such observations might lead to speculation about whether these stupendous mountains really had been 'torn and convulsed' by massive forces at some point in the past, or really were of volcanic origin (as the term 'conoid' implies). Yet, Pennant does not go beyond these implied suggestions. As we will see, however, there are other passages in Pennant's second *Tour* in particular where the sublime grandeur of Scotland's mountainous landscape coaxes him still further into wondering about the dynamic geological processes that shaped that landscape in the ancient past.

Pennant made his tours of Scotland at a time when geological debate in Britain and Europe had begun to pay increasing attention to the nature and effect of volcanoes.[18] The Neptunist geotheory, the 'standard model' of the earth's formation and development in the eighteenth century, maintained that most of the earth's rocks and topography had been formed in one or more primeval universal oceans and thus assigned only a minor role to volcanoes.[19] Although the first volume of the Comte de Buffon's *Histoire naturelle* – his *Théorie de la terre* (1749) – claimed that the earth had originally been in a molten state, it assumed that the earth had quickly cooled and that its topography had subsequently been shaped by the tidal action of a universal ocean over a very long period. Buffon thus argued against the theory of Ray and others that volcanoes are significant agents in building mountains or mountain chains, claiming that they are limited in number and caused by combustible materials lodged just below their summits rather than by a subterraneous mantel of high-tempera-ture molten material under pressure.[20] Yet, the second half of the eighteenth century witnessed an increasing attention to active and extinct volcanoes and the extent to which they had shaped, and continued to shape, the earth's geo-morphology. Porter notes that Sir William Hamilton, Patrick Brydone and Sir James Hall were just some of 'the Britons who pioneered investigation of Continental volcanoes' in the period.[21] As a British diplomat in Naples, Hamilton made observations of several eruptions of Vesuvius, including that of 1767, and a number of his accounts and drawings of Vesuvius and Etna were published in the Royal Society's *Philosophical Transactions* between 1767 and 1772 and were subsequently collected together in his *Observations on Mount Vesuvius* (1772).[22] Hamilton's publications made Vesuvius and Etna necessary viewing on the Grand Tour and helped to turn volcanoes into classic manifestations of the sublime.[23] While Hamilton brought currently active volcanoes to the attention of British naturalists, the writings of geological travellers such as Jean Étienne Guettard and Nicolas Desmarest about the Auvergne region of the French Massif Central, which began to appear in the early 1750s, presented early glimpses of the extent to which volcanic activity in the ancient past had moulded vast landscapes.[24] Even among Vulcanists, however, debates continued about the nature and cause of volcanoes. While claims that volcanoes resulted from the ignition of local and relatively superficial deposits of combustible materials seemed inadequate, the theory of subterranean heat remained speculative and difficult to explain.

The mid-century debate about the causes and effects of volcanoes overlapped with a developing controversy about the nature and origin of basalt.[25] As Laudan notes:

> We now assume that [basalt] flowed from volcanic vents in the earth's crust, in a type of vulcanism known as submarine or Hawaiian. Matters were not so clear in the late eighteenth century; neither its place in the succession, nor its mineralogy, nor its stratigraphic relations, nor evidence from contemporary volcanoes decisively indicated its origin.[26]

Various features of basalt appeared to support the dominant Neptunist theory that it had been formed by crystallization in a primeval ocean. Basaltic prismatic columns were interpreted by many mineralogists as 'enormous crystals [...] consistent with basalt's consolidation by water'.[27] In the Auvergne in 1763 and 1766, however, Desmarest linked columnar basalts and 'what appeared to be flows of basalt back to the volcanic cones. On that basis, he claimed that basalts had a volcanic origin'.[28] Desmarest's 'Mémoire sur l'origine et la nature du basalte' was not published until 1774.[29] But in 1763, Rudolf Erich Raspe's *An Introduction to the Natural History of the Terrestrial Sphere* drew on Robert Hooke's theory of subterranean heat in order to argue for the igneous origin of basalt and for the significance of volcanoes as major geological agents.[30] Yet, the classification of basalt as an igneous rock was a minority view in the 1760s; indeed, Laudan claims that 'The conviction that basalt's mineralogy indicated consolidation from water [...] [continued to seem] self-evident in the 1770s and 1780s'.[31]

These contemporary debates about volcanoes and basalt allow us to contextualize Pennant's pioneering discoveries on his tours of Scotland. As

Porter suggests, the investigation of European volcanoes in the late eighteenth century had an important impact on the interpretation of Britain's geomorphology, resulting in 'the revolutionary conception that the British Isles harboured ancient volcanoes, identifiable through landscape features and/or formations of lava, pumice, tufa and basalt'.[32] Pennant's contributions to the developing understanding of the geology of the Highlands and Islands included the tentative identification of the remains of several extinct volcanoes together with the suggestion that the Giant's Causeway and Staffa were part of a massive basaltic formation that extended all the way to Skye.

In *A Tour in Scotland, 1769*, Pennant's attention to possible volcanic features in the landscape is limited to a few instances. He tells us that he rode from Fort Augustus

> to the castle of *Tor-down*, a rock two miles west of Fort *Augustus*: on the summit is an ancient fortress. The face of this rock is a precipice; on the accessible side is a strong dyke of loose stones; above that a ditch, and a little higher a terrass supported by stones: on the top a small oval area, hollow in the middle: round this area, for the depth of near twelve feet, are a quantity of stones strangely cemented with almost vitrified matter, and in some places quite turned into black *scoria*: the stones were generally granite mixed with a few grit-stones of a kind not found nearer the place than 40 miles. Whether this was the antient site of some forge, or whether the stones which form this fortress had been collected from the strata of some *Vulcano* (for the vestiges of such are said to have been found in the Highlands), I submit to farther enquiry.[33]

Here, Pennant's cautious empiricism mostly confines him to the precise description of present appearances and he thereby keeps within the remit of the natural history of the period. When he allows himself to speculate about whether such features indicate past volcanic activity in Scotland, he submits the question 'to farther enquiry'.

In *A Tour in Scotland, 1772*, however, Pennant enquires farther into, and gradually comes to accept, the possibility that Scotland's landscape harboured extinct volcanoes. On the northern side of the harbour of East Tarbert on the Kintyre peninsula, he tells us, 'the rocks are of a most grotesque form: vast fragments piled on each other; the faces contorted and undulated in such figures as if created by fusion of matter after some intense heat; yet did not appear to me a *lava*, or under any suspicion of having been the recrement of a *vulcano*'.[34] Pennant becomes more confident, however, about the volcanic origins of features he observes later on the 1772 tour. His description of Arran, which includes the suggestion that a feature near the summit of 'the great hill *Dunfuin*' is likely to 'have been the effect of a *vulcano*',[35] constitutes one of the first published indications that the island had been partly shaped by volcanic processes. Towards the end of the first volume, Pennant reports having explored a hill at 'Beregonium', the supposed site of an ancient city on the southern shore of Loch Etive, from which

> are dug up great quantities of different sorts of pumices, or *scoria*, of different kinds: of them, one is the *pumex cinerarius*; the other the *P. molaris* of *Linnaeus*; the last very much resembling some that Mr. *Banks* favoured me with from the island of *Iceland*. The hill is doubtless the work of a *vulcano*, of which this is not the only vestige in *North-Britain*.[36]

One of the tours of Scotland that was most visibly influenced by Pennant is Thomas Garnett's *Observations on a Tour Through the Highlands and Parts of the Western Isles of Scotland, Particularly Staffa and Icolmkill* (1800), which presents a Vulcanist interpretation of Scotland's geomorphology, frequently refers to Pennant, and repeats, almost word for word, Pennant's analysis of the hill at 'Beregonium':

> THERE is a tradition, that Beregonium was destroyed by fire from heaven. In confirmation of this tradition, or rather as a proof that the fire which destroyed it came from the earth, it may be mentioned, that a high rock near the summit of one of the hills, has evidently a volcanic appearance. In most parts of the hill are likewise dug up great quantities of different sorts of pumices, or scoriæ of different kinds, particularly the *Pumex cinerarius*, and the *Pumex molaris* of Linnæus, very similar to the Iceland pumice-stone presented to Mr. Pennant by Sir Joseph Banks. These circumstances, I think, tend strongly to prove that this hill is an extinct volcano.[37]

As we will see, Garnett's failure to acknowledge Pennant's priority in identifying this apparently volcanic feature is part of a general pattern of overlooking Pennant's path-breaking contributions to the gradual realization of the extent to which Scotland's landscape had been formed by volcanic activity in the distant past. This omission is especially notable, given that later travellers tended to repeat verbatim many of Pennant's other observations.

In the second volume of *A Tour in Scotland, 1772* (1776), Pennant suggests that the castle hill of Finehaven is an extinct volcano as well as a British antiquity.[38] He also reports his discovery of 'a considerable quantity of *lava*' on Kinnoull hill near Perth, which is 'a proof of its having been an antient *vulcano*'.[39] Several later travellers would visit this hill because of its supposed volcanic nature. Faujas de Saint-Fond's exploration of Scotland in 1784, published in his *Travels in England, Scotland, and the Hebrides* in 1799, deployed his expertise on volcanic geomorphology, derived from extensive explorations of the French Massif Central in the 1770s, to interpret the landscape of the Highlands and Islands as almost entirely made up of the products of ancient volcanoes and volcanic activity. Towards the end of his tour, Faujas passed through Perth in order to examine 'the hill of Kinnoul', and found considerable evidence to demonstrate its volcanic origins.[40] But although Faujas's desire to examine this hill must have been stimulated by Pennant's suggestion that it was an extinct volcano, and though Faujas makes several references to Pennant, he never acknowledges Pennant's role in alerting the world to Scotland's volcanic past.

Following the publication of Pennant's *Tours*, reports of volcanic remains in Scotland started to appear in the *Philosophical Transactions of the Royal Society of London*, including Thomas West's 'An Account of a Volcanic Hill near Inverness' (1777), which makes no mention of Pennant, and Abraham Mills's 'Some Account of the Strata and Volcanic Appearances in the North of Ireland and Western Islands of Scotland' (1790), which describes volcanic features in several islands of the Inner Hebrides and makes three references to Pennant's 1772 *Tour* without mentioning its pioneering accounts of volcanic remains in Scotland.[41] Similarly, John MacCulloch's comprehensive survey of the 'trap islands' in his *A Description of the Western Islands of Scotland* (1819) barely mentions the names of any forerunners.[42] And, as we will see, Sir Archibald Geikie's *The Ancient Volcanoes of Great Britain* (1897) assigns only a minor role to Pennant.

The gathering evidence of Scotland's volcanic past was not uncontested. Volcanic interpretations of Scotland's geomorphology were anathema to Edinburgh's Neptunist natural historians. The Revd Dr John Walker, Professor of Natural History at the University of Edinburgh, and Professor Robert Jameson, Walker's student and successor, steadfastly denied the existence of ancient volcanoes in Scotland and would not have accepted Mills's pronouncement in the 1790 *Philosophical Transactions* that 'columnar basalts [...] are now, by almost universal consent, acknowledged to be of volcanic origin'.[43] In his lectures on geology, given at the University of Edinburgh between 1782 and 1800, Walker insisted that

> In Scotland there are few if any Earthquakes, and those that have been observed are of no importance. Therefore we cannot expect Volcanoes here, any more than we can expect a Volcanic Earthquake; and for this reason, that we have no volcanic or [...] inflammable materials for breeding or feeding Volcanoes. In 1700 the most remarkable Earthquake that we have any account of happened in Scotland. It was strongly felt from one extremity of Fife to the other, even from the Promontory of Fifeness to Dunfermline. But at this time the Coal works of Dysart in Fife were then on fire, which raged with great fierceness; and I imagine, that the appearance of this Earthquake was from that cause.[44]

As this indicates, Walker's lectures rejected the theory of subterraneous heat and treated volcanoes as relatively minor shapers of geomorphology, limited to known volcanic areas such as southern Italy and caused by local and superficial chemical or mineral deposits. In his *Mineralogy of the Scottish Isles* (1800), Jameson often belittles and dismisses the volcanic features identified by Pennant and Faujas. He also rejects Garnett's interpretation of the hill at 'Beregonium':

> At Boregonium [*sic*], which is a few miles from Dunstaffnage, there are, according to Dr. Garnet, undoubted volcanic appearances. Dr Walker informs me, that the pumice, which Dr. Garnet mentions, is the scorine from the iron furnaces, which were worked at that place by our ancestors.[45]

When Pennant pointed out the resemblance between specimens of basalt from 'Beregonium' on the west coast of mainland Scotland and Banks's Icelandic specimens, he was of course unaware of the possibility that these volcanic rocks might have been produced by intense volcanic activity that took place between 65 and 23 million years earlier which encompassed Greenland and Iceland as well as northern Ireland, the west coast of Scotland and the Inner Hebrides.[46] Yet, by this point in his second tour Pennant had indeed envisaged that a vast basaltic structure, mostly hidden under the sea, extended from the north coast of Ireland all the way to Skye.

Pennant's most spectacular volcanic 'discovery' in Scotland began with his recognition of the basaltic nature of the island of Staffa off the western coast of Mull, which he describes as 'a new giant's causeway, rising amidst the waves; but with columns of double the height of that in *Ireland*; glossy and resplendent, from the beams of the Eastern sun'.[47] In suggesting that Pennant was the 'discoverer' of Staffa's basaltic nature I am, of course, modifying the generally accepted view that Pennant was merely a conduit for relaying Sir Joseph Banks's account of his 'discovery' of the island to the outside world. My use of scare quotes acknowledges the fact that Staffa had long been known to local people and was, indeed, inhabited by a small family when Banks visited the island. It was also probably known to the monks of Iona and had been first mentioned in print in George Buchanan's *History of Scotland* (1582).[48] It is the case, however, that Staffa was overlooked by virtually all the naturalists who visited the Inner Hebrides in the following two hundred years, including Martin Martin, Richard Pococke and John Walker.

The geological wonders of Staffa thus remained unknown, at least to the outside world, when Pennant and Banks made their separate sea voyages northward along the western coast of Scotland in the summer of 1772. Banks's Iceland journal reveals that he explored Staffa on 12 and 13 August

Figure 8.1 Moses Griffith, 'an accurate view' of Staffa's 'Eastern side' (detail), from *A Tour in Scotland, and Voyage to the Hebrides, MDCCLXXII* (Chester, 1774).

1772, on his way to investigate Iceland's volcanic landscape.[49] But Pennant had sailed past Staffa and noted its basaltic pillars a month earlier on 11 July.[50] Pennant had not been able to land on the island because of 'rocky seas', but he got Moses Griffith, his servant and artist, to produce 'an accurate view [...] of its Eastern side',[51] which shows the basaltic pillars to the south of the island, but not Fingal's Cave, which Pennant does not mention (Fig. 8.1).[52] Although Pennant sighted Staffa before Banks, his role in making Staffa into one of Scotland's talismanic places for Romantic travellers and geologists was therefore limited to inserting into his *Tour* 'a most accurate account [...] copied from Mr Banks' Journal',[53] along with copies of iconic illustrations by Banks's artist, John Cleveley (Fig. 8.2).

While Pennant was the first naturalist to have sighted Staffa, it is notable that he does not contest Banks's claim to priority, suggesting that the observation of the natural history of Scotland was more important for him than elevating his own status as a pathbreaking observer or intrepid traveller. Indeed, the inclusion of Banks's account of Staffa into *A Tour in Scotland, and Voyage to the Hebrides, MDCCLXXII* is simply the most prominent example of Pennant's openness to other voices and perspectives.

After expressing his enthusiasm for Staffa's aesthetic qualities and Ossianic associations, Banks went on 'to describe it and its productions more philosophically'.[54] But although he notes that 'The stone of which the pillars are formed, is a coarse kind of Basaltes, very much resembling the Giant's causeway in Ireland',[55] and describes the exposed stratum below the pillars as 'composed of a thousand heterogeneous parts, which together have very much

Figure 8.2 Copy of John Cleveley (or John Miller), *Fingal's Cave in Staffa*, from *A Tour in Scotland, and Voyage to the Hebrides, MDCCLXXII* (Chester, 1774).[56]

the appearance of a Lava',[57] Banks does not explicitly conclude that volcanic action had played some part in the formation of Staffa and Fingal's Cave, despite the fact that his account of Iceland later in his journal reveals his expertise in the analysis of volcanic formations and phenomena.[58] Pennant adds a footnote to confirm that Staffa as a whole 'is a genuine mass of Basaltes',[59] but likewise avoids questions of causation or history, perhaps because the origin and nature of basalt remained unclear in the early 1770s.

If Pennant was disappointed not to be able to make his own philosophical observations of Staffa, he was perhaps compensated by discoveries on Skye that enabled him to posit the existence of a gigantic geological formation that appeared to link Skye to Ireland. On his way to Skye, Pennant sighted 'the rock Humbla, formed of Basaltic columns', which he informs us in a footnote 'was discovered by Mr Murdock Mackenzie'.[60] On Skye, Pennant encountered what he took to be the final link in the chain:

> Visit a high hill, called *Briis-mhawl*, about a mile South of *Talyskir*, having in the front a fine series of genuine basaltic columns, resembling the *Giant*'s causeway: the pillars were above twenty feet high, consisting of four, five and six

angles, but mostly of five: the columns less frequently jointed than those of the *Irish*; the joints being at great and unequal distances, but the majority are entire [...] The stratum that rested on this colonnade was very irregular and shattery, yet seemed to make some effort at form. The ruins of the columns at the base made a grand appearance: these were the ruins of creation: those of *Rome*, the work of human art, seem to them but as the ruins of yesterday.

At a small distance from these, on the slope of a hill, is a tract of some roods entirely formed of the tops of several series of columns, even and close set, forming a reticulated surface of amazing beauty and curiosity. This is the most northern *Basaltes* I am acquainted with; the last four in the *British* dominions, all running from South to North, nearly in a meridian: the *Giant's Causeway* appears first; *Staffa* succeeds; the rock *Humbla* about twenty leagues further, and finally the column of *Briis-mhawl*: the depth of ocean in all probability conceals the lost links of this chain.[61]

By suggesting that the ruins of the columns at the base of the geological formation he discovered at 'Briis-mhawl' 'were the ruins of creation', in comparison with which 'those of *Rome*, the work of human art, seem [...] but as the ruins of yesterday', Pennant indicates both their grand appearance and the long timescale involved in the creation and destruction of this feature. But Pennant is also implicitly suggesting that his discovery is comparable to Banks's discovery of Staffa. As Allison Ksiazkiewicz shows in this volume (Chapter 9), Banks began a long-standing trend by comparing the formations of basaltic pillars on Staffa to architectural features and insisting that this natural architecture exceeds 'the cathedrals or the palaces built by men' and even the achievements of ancient Greece.[62] Indeed, Pennant's apparently modest account of his discovery of what he takes to be a massive geological formation that extends from the Giant's Causeway to the south-west coast of Skye, would appear to outdo Banks. Once again, however, he plays down the potential sublimity of these physical and temporal dimensions and does not speculate about the dynamic forces that might have produced such an enormous geological structure.

Pennant had indeed identified part of a gigantic geological formation that was even more extensive than he imagined and would gradually be reconstructed by geological travellers over the following hundred years or so. The bulk of the second volume of Sir Archibald Geikie's *The Ancient Volcanoes of Great Britain* (1897) is given over to 'The Volcanoes of Tertiary Time' (now known as the Palaeogene period, when the Atlantic Ocean began to open up, producing major volcanic activity along the rift). Although Tertiary-period eruptions occurred 'Over many regions of the European continent', in the north-west they assumed more colossal proportions, and took the form of fissure-eruptions by which many thousands of square miles of country were deluged with lava. From the South of Antrim all along the West of Scotland to the north of the Inner Hebrides remains of these basalt-floods form striking features in the existing scenery. The same kind of rocks reappear in the Faroe Islands and in Iceland, so that an enormous tracts of North-western Europe, much of it now submerged under the sea, was the scene of activity of the Tertiary volcanoes. In entering, therefore, upon a consideration of the British Tertiary volcanic rocks, we are brought face to face with the records of the most stupendous succession of volcanic phenomena in the whole geological history of Europe. Fortunately these records have been fully preserved in the British Isles, so that ample materials remain there for the elucidation of this last and most marvellous of all the volcanic epochs in the evolution of the continent.[63]

Because the volcanic remains of the Inner Hebrides are more obvious than those from earlier periods and in other places, their investigation, beginning in the late eighteenth century, was an important stage in the emerging understanding of volcanic rocks and topography. Geikie thus presents a history of the investigation of these volcanic formations. He restricts Pennant's contribution to the understanding of volcanism in Scotland to relaying Banks's account of Staffa and Fingal's Cave to the outside world.[64] The earliest important contribution, in Geikie's view, was made by John Whitehurst in the second edition of *An Inquiry into the Original State and Formation of the Earth* of 1786, 'who gave a good account of the basalt cliffs of Antrim, and regarded the basaltic rocks as the result of successive outflows of lava from some centre now submerged beneath the Atlantic'.[65] More developed observations appeared in Mills's 'Some Account of the Strata and Volcanic Appearances in the North of Ireland and Western Islands of Scotland' (1790) and in Faujas de Saint-Fond's *Travels* (1799), which 'may be taken as the beginning of the voluminous geological literature which has since gathered round the subject'.[66] But Geikie gives pride of place to MacCulloch's *Description of the Western Islands of Scotland* (1819), which, as noted earlier, hardly mentions any predecessors.

  Pennant's pioneering contribution to the reconstruction of the volcanic formations that extend from the north of Ireland to Skye and beyond was therefore largely ignored or overlooked by subsequent geological travellers and historians. Yet, the implications of Pennant's finds were not missed by the leaders of the Neptunist school in late eighteenth-century Edinburgh. When Jameson toured Skye in 1798 he refers rather dismissively to Pennant's basaltic formation on 'Briis-mhawl': 'at little Breeze hill, which is near to the vale of Talysker, there is a pretty colonade of basalt pillars, which, Mr Pennant, in his voyage to the Hebrides, erroneously mentions as the most northern groupe of columns in Scotland'.[67] By 1798, as we have seen, the volcanic interpretation of basalt had become a serious challenge to the Neptunist geotheory, which may account for Jameson's dismissive tone here. But when Jameson later discovered his own impressive columnar formation (of volcanic pitchstone rather than basalt), the Sgurr of Eigg, which perhaps compensated for his own failure to visit Staffa, he ironically extended the colossal formation that Pennant had envisaged (though he did not associate it with volcanism). As D. R. Oldroyd and B. M. Hamilton point out, Archibald Geikie's investigation of Eigg in 1864 'suggested that the basaltic lavas of the Hebrides, parts of the mainland, and also across in Northern Island, might be mere fragments of what was formerly one vast tract of basaltic lavas'.[68] Geikie does not, however, acknowledge Pennant's much earlier vision of this geological formation.

Although Pennant was one of Britain's foremost natural historians, who set out to write pioneering natural history tours of the Highlands and Islands of Scotland, what he encountered and discovered there triggered the beginnings of a paradigm shift. Pennant's mode of scientific observation, self-representation and writing style in his *Tours* of Scotland are largely confined within the conventions of eighteenth-century natural history. His outlook on his journeys through the Highlands and Islands mostly avoids speculations about past and future, cause and effect, in favour of observations and descriptions of static features. Yet, Pennant made observations of geomorphic features that can be regarded as pioneering fieldwork for the geological exploration of Scotland in later decades. His identification of a range of topographical features that he believed were the remnants of ancient volcanoes inevitably implied that Scotland's landscape had been shaped by dynamic forces in the past. Pennant does not pursue that line of thought very far, but his fieldwork in Scotland nonetheless contributed to 'the revolutionary concept that the British Isles harboured ancient volcanoes', and this in turn contributed to Romantic geology's view of the earth as a dynamic system that had been active for an enormously long time.[69] If the history of earth science in the eighteenth century can be plotted in terms of the transition from mineralogy to geology, as both Porter and Laudan suggest, then Pennant's tours of Scotland helped pave the way for that transition and took a few hesitant steps along it. The careful observation of natural features in Pennant's natural history tours of Scotland went some way towards challenging some of the fundamental assumptions of natural history.

Notes

* The research and drafting of this chapter were made possible by a Leverhulme Trust Research Fellowship for the academic year 2013–14.

1 Brian D. Osborne, introduction to Thomas Pennant, *A Tour in Scotland, 1769* (Edinburgh: Birlinn, 2000), xi, xviii. On the eighteenth-century 'discovery' of Scotland

by writers and artists, see James Holloway and Lindsay Errington, *The Discovery of Scotland: The Appreciation of Scottish Scenery through Two Centuries of Painting* (Edinburgh: The National Gallery of Scotland, 1978); John Glendening, *The High Road: Romantic Tourism, Scotland, and Literature, 1720–1820* (New York: St Martin's Press, 1997); Alastair J. Durie, *Scotland for the Holidays: Tourism in Scotland, c.1780–1939* (East Linton: Tuckwell Press, 2003); Katherine Haldane Grenier, *Tourism and Identity in Scotland: Creating Caledonia* (Aldershot and Burlington: Ashgate, 2005). On Pennant's responses to landscape in Scotland, see Paul Smethurst, 'Peripheral Vision, Landscape, and Nation-Building in Thomas Pennant's Tours of Scotland, 1769–72', in *Travel Writing and Tourism in Britain and Ireland*, ed. Benjamin Colbert (London: Palgrave Macmillan, 2011), 13–30.

2 Charles W. J. Withers, introduction to Thomas Pennant, *A Tour in Scotland and Voyage to the Hebrides, 1772*, ed. Andrew Simmons (Edinburgh: Birlinn, 1998), xx.

3 Thomas Pennant, *Tour on the Continent, 1765, by Thomas Pennant, Esq.*, ed. G. R. de Beer (London: Ray Society, 1948), vi.

4 Charles W. J. Withers, ODNB *s.n.* Pennant, Thomas (1726–1798) (accessed 27 April 2014).

5 See Rachel Laudan, *From Mineralogy to Geology: The Foundations of a Science* (Chicago and London: University of Chicago Press, 1987); David Oldroyd, *Thinking about the Earth: A History of Ideas in Geology* (London: Athlone, 1996); Gordon L. Davies, *The Earth in Decay: A History of British Geomorphology, 1578–1878* (London: Macdonald Technical and Scientific, 1969), 279; Roy Porter, *The Making of Geology: Earth Science in Britain, 1660–1815* (Cambridge and New York: Cambridge University Press, 1977), 102, 114, 120; Martin J. S. Rudwick, *Bursting the Limits of Time: The Reconstruction of Geohistory in the Age of Revolution* (Chicago and London: University of Chicago Press, 2005), 77; Dennis R. Dean, *James Hutton and the History of Geology* (Ithaca and London: Cornell University, 1992), 13.

6 See Porter, *Making of Geology*, 112–14. For Bacon's vision of an all-encompassing scientific project, including natural history, see Francis Bacon, *The New Organon*, ed. Lisa Jardine and Michael Silverthorne (Cambridge and New York: Cambridge University Press, 2000), especially the concluding 'Outline of a Natural and Experimental History' (222–38).

7 Porter, *Making of Geology*, 32–41; Porter is quoting Hans Sloan and Edward Lhuyd.

8 Ibid. 38–41.

9 Ibid. 112–14, 120–27.

10 Laudan, *From Mineralogy to Geology*, 70–102; on eighteenth-century natural history and mineralogy, see Rudwick, *Bursting the Limits of Time*, 37–48, 59–71.

11 Porter, *Making of Geology*, 170–76.

12 See Anthony Hallam, 'Neptunists, Vulcanists, and Plutonists', *Great Geological Controversies* (Oxford and New York: Oxford University Press, 1983), 1–28.

13 Porter, *Making of Geology*, 218.

14 See Carl Thompson, *The Suffering Traveller and the Romantic Imagination* (Oxford and New York: Clarendon Press, 2007).

15 Thomas Pennant, *A Tour in Scotland, and Voyage to the Hebrides, MDCCLXXII*, 2 vols (Chester, 1774), I: 56.

16 Idem, *A Tour in Scotland, 1769* (Chester, 1771), 52.

17 Idem, *A Tour in Scotland [...] 1772*, I: 364–65.

18 See Porter, *Making of Geology*, 160–65; Laudan, *From Mineralogy to Geology*, 181–93.

19 For Rudwick's account of the standard model, see *Bursting the Limits of Time*, 172–80.

20 See Comte de Buffon, *Natural History, General and Particular*, volume I, trans. William Smellie (Edinburgh, 1785), Article XVI: 408–41. As Laudan, *From Mineralogy to Geology*, 69, notes, however: 'By 1778, when he published the *Époques de la nature*, Buffon had come to the conclusion that heat played a major role in the late as well as early stages of earth history'.

21 Porter, *Making of Geology*, 163.

22 William Hamilton, *Observations on Mount Vesuvius, Mount Etna, and other Volcanos: in a Series of Letters, Addressed to the Royal Society* (London, 1772).

23 Geoffrey V. Morson, ODNB *s.n.* Hamilton, Sir William (1731–1803) (accessed 4 May 2014):

> In 1776 Hamilton published his major work on vulcanology (with text in both English and French), Campi phlegraei: Observations on the Volcanoes of the Two Sicilies; it included more than fifty spectacular hand-coloured gouache illustrations by Pietro Fabris of eruptions, lightning, and other natural phenomena. This publication did a great deal to make volcanoes – the 'fields of fire' alluded to in the title – a popular subject in art and poetry and to cause a visit to Vesuvius to be a necessary stage on the grand tour.

24 See K. L. Taylor, 'Geological Travellers in Auvergne, 1751–1800', in *Four Centuries of Geological Travel: The Search for Knowledge on Foot, Bicycle, Sledge and Camel*, ed. P. N. Wyse Jackson (London: Geological Society, 2007), 73–96. Also see Hallam, *Great Geological Controversies*, 6–9; Rudwick, *Bursting the Limits of Time*, 203–12.

25 On the basalt controversy, see Rhoda Rappaport, 'The Earth Sciences', and Charlotte Klonk, 'Science, Art, and the Representation of the Natural World', in *The Cambridge History of Science, Volume 4: Eighteenth-Century Science*, ed. Roy Porter (Cambridge and New York: Cambridge University Press, 2003), 417–35 (426–28), 584–617 (606–9); see also Rudwick, *Bursting the Limits of Time*, 62–63, 94, 105–8, 204–7.

26 Laudan, *Mineralogy to Geology*, 181.

27 Ibid. 182.

28 Ibid. 183.

29 See Nicolas Desmarest, 'Mémoire sur l'origin et la nature du basalte', *Mémoires de l'Académie Royale des Sciences* (Paris, 1774), 705–75.

30 Laudan, *Mineralogy to Geology*, 184; see Rudolf Erich Raspe, *An Introduction to the Natural History of the Terrestrial Sphere*, trans. and ed. A. N. Iversen and A. V. Carozzi (New York: Hafner, 1970).

31 Laudan, *Mineralogy to Geology*, 182.

32 Porter, *Making of Geology*, 162.

33 Pennant, *A Tour in Scotland, 1769*, 173–74.

34 Idem, *A Tour in Scotland […] 1772*, I: 188.

35 Ibid. I: 211.

36 Ibid. I: 412. See http://canmore.rcahms.gov.uk/en/site/23247/details/beregonium+benderloch (accessed 27 April 2014):

> The Pictish city of Beregonium is said to have been situated between Dun Mac Sniachan and Dun Bhaile an Righe […] The RCAHMS [Royal Commission on the Ancient and Historical Monuments of Scotland] states that the spurious name 'Beregonium', a mis-reading of Ptolemy's 'Rerigonium', was mistakenly applied by Hector Boece in his 'Scotorum Historiae'.

37 Thomas Garnett, *Observations on a Tour through the Highlands and Parts of the Western Isles of Scotland, Particularly Staffa and Icolmkill*, 2 vols (London, 1800), I: 279–80.

38 Thomas Pennant, *A Tour in Scotland, MDCCLXXII*, 2 vols (London, 1776) II: 165–66. Shortly after the publication of this volume, John Williams, in his *An Account of Some Remarkable Ancient Ruins, Lately Discovered in the Highlands, and Northern Parts of Scotland* (Edinburgh, 1777), identified a vitrified fort on the summit of 'Finaven' (39–40). Williams's general account of vitrified forts in the Highlands offered an alternative way of interpreting some of Pennant's volcanoes.

39 Pennant, *A Tour in Scotland [...] 1772*, II: 118.

40 See B. Faujas de Saint-Fond, *Travels in England, Scotland, and the Hebrides*, 2 vols (London, 1799), II: 185.

41 See Thomas West, 'An Account of a Volcanic Hill near Inverness. In a Letter from Thomas West, Esq. to Mr Lane, F. R. S.', *Philosophical Transactions of the Royal Society of London* LXVII (1777): 385–87; Abraham Mills, 'Some Account of the Strata and Volcanic Appearances in the North of Ireland and Western Islands of Scotland. In two Letters from Abraham Mills, Esq. to John Lloyd, Esq. F. R. S.', *Philosophical Transactions of the Royal Society of London* LXXX (1790): 73–100.

42 See John MacCulloch, *A Description of the Western Islands of Scotland, Including the Isle of Man: Comprising an Account of their Geological Structure; with Remarks on their Agriculture, Scenery, and Antiquities*, 3 vols (London and Edinburgh, 1819), I: 235–587; II: 1–79.

43 Mills, 'Some Account', 88.

44 John Walker, *Lectures on Geology, Including Hydrography, Mineralogy, and Meteorology with an Introduction to Biology*, ed. Harold W. Scott (Chicago and London: University of Chicago Press, 1966), 211–12.

45 Robert Jameson, *Mineralogy of the Scottish Isles; with Mineralogical Observations Made in a Tour through Different Parts of the Mainland of Scotland* (Edinburgh, 1800) I: 197n.

46 See Alan McKirdy, John Gordon and Roger Crofts, *Land of Mountain and Flood: The Geology and Landforms of Scotland* (Edinburgh: Birlinn, 2009), 150–56.

47 Pennant, *Tour in Scotland [...] 1772*, I: 298. The large pyramidal columns of basalt that made up the Giant's Causeway had long been the subject of debate by British naturalists; see Porter, *Making of Geology*, 40, 113.

48 For surmises about the local islanders' and Iona's monks' knowledge of Staffa, see Donald B. MacCulloch, *The Wondrous Isle of Staffa* (Edinburgh and London: Oliver and Boyd, 1957), 5–7.

49 See Roy A. Rauschenberg, 'The Journals of Joseph Banks's Voyage up Great Britain's West Coast to Iceland and to the Orkney Isles, July to October, 1772', *Proceedings of the American Philosophical Society* 117 (1973): 186–226 (206).

50 Pennant, *Tour in Scotland [...] 1772*, I: 298.

51 Ibid. I: 299.

52 Withers, *Tour in Scotland, 1772*, xvi. See Donald Moore, *Moses Griffith, 1747–1819: Artist and Illustrator in the Service of Thomas Pennant* (Caernarfon, 1979).

53 Pennant, *Tour in Scotland [...] 1772*, I: 299–309.

54 Ibid. I: 302.

55 Ibid. I: 308–9.

56 This iconic image of Fingal's Cave seems to have been made by James Miller, not John Cleveley, as is often claimed. See Allison Ksiazkiewicz's essay in this volume.

57 Ibid. I: 306.

58 For Banks's description of the volcanic nature of Iceland, see Rauschenberg, 'Journals of Joseph Banks's Voyage': 214–25.

59 Pennant, *Tour in Scotland [...] 1772*, I: 309.

60 Ibid. I: 310. See Murdoch Mackenzie, *Nautical Descriptions of the West Coast of Great Britain, from Bristol Channel to Cape-Wrath* (London, 1776).

61 Pennant, *Tour in Scotland [...] 1772*, I: 334–35.

62 Ibid. 301. See also the essay by Allison Ksiazkiewicz in this volume (Chapter 9).

63 Sir Archibald Geikie, *The Ancient Volcanoes of Great Britain*, 2 vols (London: Macmillan, 1897), II: 108. For modern geological interpretations of the vulcanism of the Inner Hebrides, see J. B. Whittow, *Geology and Scenery in Scotland* (Harmondsworth: Penguin, 1977), chapters 11–13; Con Gillen, *Geology and the Landscapes of Scotland* (Harpenden, Hertfordshire: Terra Publishing, 2003), chapter 7; McKirdy, Gordon and Crofts, *Land of Mountain and Flood*, 151–54.

64 Geikie, *Ancient Volcanoes*, II: 109

65 Ibid. See John Whitehurst, *An Inquiry into the Original State and Formation of the Earth*, 2nd ed. (London, 1786).

66 Geikie, *Ancient Volcanoes*, II: 109.

67 Jameson, *Mineralogy of the Scottish Isles*, II: 73–74.

68 D. R. Oldroyd and B. M. Hamilton, 'Themes in the Early History of Scottish Geology', in *The Geology of Scotland*, ed. Nigel H. Trewin (London: The Geological Society, 2002), 27–43 (40). See Archibald Geikie, *The Scenery of Scotland Viewed in Connection with its Physical Geology*, 3rd ed. (London: Macmillan, 1901), 141–44, 167–73.

69 Porter, *Making of Geology*, 163.

Chapter 9

GEOLOGICAL LANDSCAPE AS ANTIQUARIAN RUIN: BANKS, PENNANT AND THE ISLE OF STAFFA

Allison Ksiazkiewicz[*]

Scholars investigating the entwined nature of art and science in the late eighteenth century have rightfully noted and remarked upon the architectural vocabulary that Joseph Banks (1743–1820) used in his pictorial and textual description of the island of Staffa, as published in Thomas Pennant's *A Tour in Scotland, and Voyage to the Hebrides, MDCCLXXII* (1774).[1] References to both Gothic and Classical architectures conveyed the visual affect of regularity or the appearance of artifice that the basaltic columns displayed.[2] It presented, as Banks claimed, 'a very singular sight'.[3] Understandably, scholarship regularly frames Staffa as a geological object. Banks was, after all, on his way to Iceland to study volcanic activity and botanize;[4] however, the account published in *A Tour in Scotland* reflects a limited engagement with current debates in the geological or mineralogical sciences.[5] In the context of the *Tour*, Banks's description of Staffa could be better understood as an antiquarian object of study, one which resonates well with the overlapping interests of naturalists and antiquarians, and engages a picturesque vision of the British landscape.[6]

As the account is fixed between antiquarian descriptions of Iona's Gothic cathedral and Cairn na Burgh More's ancient fortress, the research programmes of dilettanti and antiquarian groups provide a rich and nuanced framework for thinking about Staffa, especially since both Thomas Pennant and Joseph Banks were enthusiastic associates of antiquarian circles. From 1754 to 1760, Pennant was a member of the Society of Antiquaries of London,[7] and his extensive correspondence testifies to a continued engagement in antiquarian matters until the 1790s. His second tour in Scotland, which included excursions through the Hebrides, was well planned and organized. Queries were circulated to various parishes in anticipation of his arrival. Picturesque descriptions of landscape and antiquities played an important role in his revised and expanded account of Scotland, and supported a vision of 'improved' landscape that ideologically united Britain with its most remote and wild regions.[8] As John Bonehill has argued, Banks was also motivated by an idea of a unified British heritage and was significantly influenced by antiquarian interests in his decision to travel to the North.[9] His formal involvement in the Society of Antiquaries of London included sitting on the Council, and as member of the Society of Dilettanti he served various positions such as 'Very High Steward', Treasurer and Secretary.[10] By considering Banks's description as an extension of contemporary antiquarian and dilettanti projects, two traditional readings of Staffa are challenged. First, while the sublime character of Staffa is regularly the only aesthetic reaction discussed in scholarship, the picturesque plays a significant force in the explorers' observation practices, offering a framework for understanding the landscape and cultural assimilation of the site. Second, Staffa considered as an 'antiquarian ruin' establishes the conceptual connection between natural history and aesthetics in concrete terms. The aesthetic sensibilities of dilettanti and antiquarian groups, particularly the picturesque, directly informed how Banks approached the problem of describing an unknown landscape or set of ruins that effectively assimilated discreet and individual elements into a vision of a unified whole. In this instance, a history of the arts and a scientific interest in human and natural history conflate in the making of Staffa's cultural landscape, where aesthetic sensibility shaped the behaviour of scientific explorers.

After withdrawing from the *Resolution* voyage to the South Pacific over disagreements regarding shipboard facilities, Sir Joseph Banks decided to sail north to Iceland, as the first foreign naturalist to undertake an expedition in that country.[11] While anchored at the Sound of Mull in the Scottish Hebrides for provisions and to prepare a visit to Iona,[12] Banks decided to oblige local hospitality and joined Sir Allan Maclean at Drimnin House for breakfast one morning. There, an English gentleman named Mr Leach reported to them that a nearby island had 'pillars like those of the Giants Causeway'.[13] The Irish example was known to Banks, as were examples of columnar basalt in Germany and France. Since 1751, an increasingly heated debate had been underway within scientific circles regarding the classification of basaltic rocks. One group claimed that the columns originated from a watery matrix as a mudstone or hardened sediment rock. The other group claimed that basalt was a type of volcanic lava.[14] The Royal Society published and read several papers by naturalists and philosophers on the aqueous or volcanic origins of basalt. Prints of the Giant's Causeway were included as a visual reference for Richard Pococke's articles about its prismatic nature and functioned as proxy images for naturalists such as Nicolas Desmarest who were unable to visit the Irish site.[15] Banks was aware that columnar basalt formations had been recorded in nearby Canna, Skye and Mull, but added that 'they seemd, from the descriptions I had heard or read of them, to be little more than Efforts of nature towards that regularity, which, in the giants causeway, she had completed'.[16] The Irish Causeway was the most impressive example of these natural formations in the North, and since the season was too late to include the Causeway on Banks's itinerary as initially planned, 'this was an opportunity not to be lost'.[17] That afternoon he set out with a small boat-party guided by Mr Leach and the son of Mr Maclean. By the time they arrived at Staffa it was past nightfall. They pitched camp, ate dinner, and spent the remainder of the evening listening to their guides describe the island with enthusiasm.[18] The following morning, the group was anxious to begin exploring.

As an object of antiquarian curiosity, Staffa occupied an intermediate position between antiquarian studies of the past and dilettanti programmes for reinventing the present. Provoked by the threat of a disappearing Highland culture in the wake of Culloden and a changing land economy, antiquarians recorded ancient monuments and collected oral histories in order to document and preserve fading indigenous tradition.[19] In the case of remote regions such as northern Scotland, the disappearing culture was of British significance and fuelled a burgeoning sense of 'national' heritage. The publication of James MacPherson's Ossianic poetry perhaps did the most to raise awareness of ancient Highland culture within Britain and Europe.[20] The Ossianic poems described a 'golden age' of Scotland's ancient past, and testified to the prowess of the Scottish 'noble savage' from whom the modern Highland people were believed to be directly descended.[21] Often compared to the Homeric epics such as the *Iliad*,[22] the Ossianic legends described a remote period of Highland history and the heroic virtues of ancient warriors. MacPherson's published works fuelled the reading public's imagination, including that of Banks. 'To have read ten pages of Ossian under the shades of those woods would have been a luxury above the reach of kings', wrote Banks as he sailed the Sound of Mull past the hills of Morvern, 'the mother of the romantick scenery of Ossion'.[23] For Banks and later travellers, Staffa functioned as a monument to Ossianic lore. 'How fortunate that in this cave we should meet with the remembrance of that chief, whose existence, as well as that of the whole *Epic* poem, is almost doubted in *England*',[24] wrote Banks. Within this context, Staffa became an artefact that affirmed a singular heritage for a newly unified Britain and operated within the antiquarian paradigm in which objects rendered text relevant.[25]

Antiquarians were not the only group reexamining ancestral sources of culture. In search of novel models for art and architecture that exemplified a liberal nation, the Society of Dilettanti promoted the study of ruins beyond the established canon at Rome. Representing an old cultural regime of prescribed conventions, the examples in Rome were no longer a viable source of learning or politic.[26] In turn, ancient Greece came to symbolize republican virtues that the Dilettanti supported, and its ruined architectures presented new types for contemporary design in Britain. As a result, the ruins in Rome were reimagined as poor derivatives of their original Grecian models: 'But Athens, the mother of Elegance and Politeness, whose magnificence of scale yield to that of Rome, and who for the beauties of a correct style must be allowed to surpass hers as much as an original excels a copy.'[27] While antiquarian and dilettanti groups treated the past differently for their respective aims, there was a growing awareness and interest in destabilizing established sources of culture in terms of geography, politics and identity.

Banks's description of Staffa, in particular the allusions to Gothic and Classical arts, engaged the sensibilities of both antiquarian and dilettanti groups; however, the aesthetic quality of each style was expressed in different ways in his account.[28] Recalling William Gilpin's early writing on the picturesque, Bank's description 'painted' Staffa 'in a double light, with regard to the whole, and with regard to its *parts*'.[29] Considering 'the whole before its parts, as it naturally precedes in practice',[30] Banks portrayed the general form of the island as a Classical architectural-ruin.[31] In particular the south side of the island, where the basalt pillars and tufa stratum reached its tallest point, best moulded into an image of a Classical ruin as described by dilettanti in their accounts of ancient antiquities from the Greek and Roman empires. An association between Classical ruins and Staffa's natural formations was re-enforced by the accompanying pen-and-ink drawings produced by one of the expedition's draughtsmen, John Frederick Miller (*fl.* 1768–96). The south-west view as depicted by Miller suggests the overall appearance of a triangular pediment supported by stone columns. The tallest point of the island, where the height of the pillars and the thickness of the stratum exceed anywhere else on Staffa, features at the centre of the composition.[32] From this angle on the water, the vertical basalt-pillars and triangular tufa might suggest to the observer a Grecian colonnade supporting a ruined pediment. From a distance, the island as a whole was described and drawn by Banks and Miller as a curious landscape that possessed characteristics reminiscent of Grecian architecture.

The mode of architectural analogy changes dramatically when Banks and his crew step onto the island and begin to explore its individual parts such as Fingal's Cave. The attention to localized particulars recalls a vocabulary of Gothic cathedrals, in which the effect of light on the surface of the basalt and the feeling of movement created by the repetition of shapes dominate the description. Banks's account traces the procession of vertical columns down the interior of the space and follows the motion of his eye to the ceiling composed of the overhanging remnants of broken pillars. The cave, he observes, is airy and lighted from without so that one was able to observe its entire length. Yellow crystallized limestone had seeped between the pillars, setting off the dark basaltic rock in a play between light and dark materials. The bright colour of the calcareous sparry-matter defined the angles of the columns with 'elegance', and produced a visual variety that was 'still more agreeable'. From the ceiling and walls, the viewer's attention is drawn down to the cave floor which, covered beneath 12 feet of water, is a constant flux of motion as the tide enters and exits the space.[33] The cave's Gothic character is further empha-sized by the suggestion of a vaulted ceiling in the pen-and-ink drawing attrib-uted to James Miller (*fl.* 1770s), John Frederick's younger brother, who also accompanied Banks on his voyage to the North (Fig. 9.1).

The layered two-part structure of Banks's account effectively combines the mixed styles of the 'ruin' into a single entity that, in a picturesque manner, makes reference to antiquarian and dilettanti subjects. Attention to empirical form enhanced by associations with the ancient arts produces a 'picture' of a pseudo-architecture, or a natural form that suggests an architectural proto-type. Like all prototypes or original sources discovered during the eighteenth century, acquiring an accurate record of the artefact or architecture was para-mount. Collecting the exact measurements of ancient remains was a shared

Figure 9.1 James Miller, *Fingal's Cave, Staffa* (1772). Topographical drawings from Banks's voyage to Iceland, pen and ink on paper.

practice for both antiquarians and dilettanti, despite their different goals. The former measured and recorded local ruins and artefacts with the aim of observing and improving historical accuracy, as a means of promoting a lost British heritage. The latter performed the same activities abroad in order to interpret and manipulate ancient sources for the improvement of taste.

While the Society of Antiquaries contemplated British artefacts and history, the Society of Dilettanti focused on the Classical world of the Mediterranean, and actively sponsored members to travel and document the ancient architectures of this region.[34] The decline of the Ottoman Empire during the seventeenth century opened Greece to English travellers looking to expand their education beyond established itineraries of the Grand Tour. In 1751 the Society sponsored two gentlemen, James Stuart (1733–88) and Nicholas Revett (1720–1804) to tour Greece and publish an account of monuments of the Acropolis and the surrounding Athenian countryside.[35] The resulting publication, *Antiquities of Athens and other monuments of Greece* (1762), was in part inspired by and modelled on Antoine Desgodetz's *Les edifices antiques de Rome dessinés et mesurés très exactement* (1682), the first comprehensive survey of Roman buildings and ruins in Rome.

Published for architects, freemasons, carpenters, decorative artists and painters, *Antiquities of Athens* was produced in dialogue with *Edifices de Rome* as a pattern book for the improvement of taste.[36] The preface to the Grecian tour claimed that previous studies, such as Desgodetz's, had been inadequate for want of accurate measurements of the ruins.[37] Careful study, meticulous measurement and scrupulous delineation were the core objectives of Stuart's and Revett's project. *Antiquities of Athens* also reflected current attitudes towards architecture and marked a departure from early modern architectural theory that envisioned a building as a unified body. Baroque and Counter-Reformation architecture considered proportion and design based on optical effects and the totalizing principles of perspective; however, the eighteenth-century Grand Tourist did not engage with antique models and architectural theory in the same way. When visiting Rome, architectural students saw and experienced structures as ruins or pieces of building.[38] The increased emphasis on architectural fragments abstracted these elements from the design principles of the original structure and transformed ancient ruins into a series of individual parts that could be manipulated in new and creative ways. Elements of the antique were appropriated by the student and then made to 're-perform' as components of London buildings.[39] By contrast, *Edifices de Rome* approached architecture though the older 'holistic' tradition in its attempt to study the integrity of a building in its entirety. Degodetz's treatise began with a study of the most architecturally significant building in Rome – the Pantheon – an engineering triumph and an exemplary manifestation of perfect shapes.[40] The engravings portrayed the Pantheon as a complete building in order to study its proportions. By contrast the Doric column, considered a rude or primitive architectural element in the eighteenth century, introduced Stuart's and Revett's study of ancient Greece. In this instance, the column was treated as an isolated object intended to reference antiquity rather than represent a complete design ideology.

The playful appropriation and manipulation of parts drew attention to the surface or materiality of a structure because a visual reference and its allusion to a particular ruin depended on accurate representation. Picturesque vision embraced both the careful delineation or attention to isolated parts of a whole and their reassembly into a new kind of 'picture'.[41] As Gilpin explained: 'Nature should be the standard of imitation; and every object should be executed, as nearly as possible, in *her manner*.'[42] Reducing observation practice and visual expression to individual details generated a sense of empirical accuracy. Attention to Nature in this way aided the artist in representing 'that endless variety, which nature exhibits on every subject', according to Gilpin. He continued to suggest that those who 'deviating from this standard, instead of nature, have recourse only to their own ideas. They have gotten a general idea of a man, a horse, or a tree; and to these ideas they apply upon all occasions', thus 'a sameness runs through all their performances. Every figure, and every tree bears the same stamp'.[43] In this way, the artist projects onto the canvas from their imagination rather than engaging with and portraying Nature itself. The reunion of parts in picturesque vision produced a mode of 'seeing' that was layered with numerous textures of cultural meaning, and which regularly dislocated the object from its original context. The result was 'multi-informational' in which the processing and ordering of fragments and perspectives produced a collaged effect.[44] Perhaps this is why the role of 'harmony' in uniting individual elements becomes so important for later picturesque theory. The empirical and picturesque nature of Stuart's and Revett's project enabled readers in Britain to consume the antique through the immediacy of material and physical structure.[45]

The appropriation of ruins as reimagined and manipulated by eighteenth-century antiquaries and dilettanti also applied to natural architectural forms such as Staffa. The architectural theorist Joseph Gandy (1771–1843), who regularly drafted for architect John Soane (1753–1837), produced the watercolour *Origins of Architecture* as a frontispiece for his unpublished treatise 'The art, philosophy and science of architecture'. Fingal's Cave features on the right side of the image as a natural model of architecture. Allusions to the broken basalt columns that form the cave ceiling also appeared in Soane's Gothic ceiling for the Court of Chancery at Westminster Hall.[46] In debates about original types and the primacy of either Greek or Roman design, the example of Staffa further illustrates a widening cultural interest in reevaluating sources of invention in the arts. Uno von Troil (1746–1803), later ordained as archbishop of Uppsala, was a member of Banks's excursion party to Staffa. According to his *Letters on Iceland* (1780), one of the few major accounts of the northern voyage, Fingal's Cave was a natural monument that overshadowed one's capacity to admire human art: 'How magnificent are the remains we have of the porticos of the ancients! and with what admiration do we behold the colonnades which adorn the principal buildings of our times!' But the Louvre in Paris, St Peter's Basilica in Rome and the ruins at Palmyra all paled by example. Troil continued his comparison: 'this piece of Nature's architecture far surpasses every thing that invention, luxury, and taste ever produced among the Greeks'.[47]

Like his travelling companion Troil, Banks's admiration at the regularity and height of the basalt columns quickly gave way to a comparison between the arts and natural models of artifice; however, the tone and vocabulary of Banks's description does not so readily fall into the category of the sublime as is generally asserted in scholarship.[48] While the initial grandeur of the scene and the brief digressions into the imagination recall a sublime response, ideas of 'splendor' quickly gave way to an attention to the empirical qualities of the landscape. After being 'struck with a scene of magnificence which exceeded our expectations',[49] Banks provided a quick overview of the landscape reminiscent of a dilettanti's picturesque vision of ruins. Qualifying 'magnificence', he wrote:

> the whole of that end of the island [is] supported by ranges of natural pillars, mostly above 50 feet high, standing in natural colonnades, according as the bays or points of land formed themselves; upon a firm basis of solid unformed rock, above these, the stratum which reaches to the soil or surface of the island, varied in thickness, as the island itself formed into hills or vallies.[50]

As suggested by the justification of his reaction, Banks responded 'rationally' to the curious landscape before him and was not carried away by superfluous emotion. In this way, Banks continued to remain engaged with the scene and was not overwhelmed by his feelings regarding its strangeness and novelty, as would have been the case in a sublime register.

Pennant's own brief description of Staffa immediately precedes that of Banks and is likewise a strange mix of sublime and picturesque affects. Unable to land on the island due to rough waters, Pennant describes the island from his little boat as Mull opened up to view: 'Nearest lies Staffa, a new giant's causeway, rising amidst the waves; but with columns of double the height of that in Ireland; glossy and resplendent, from the beams of the Eastern sun.'[51] The scene is brimming with activity. Staffa is described 'emerging' from the depths of the waters as if Pennant is witnessing a geological drama unfold. The allusion to movement or growth suggests a sublime landscape. Sunlight glittering off the surface of the crystalline columns, like a gem catching the light, emphasizes the dazzling effect of this unique rock formation in the mind's eye. The descriptions before and after Pennant's initial impression are plainly picturesque in sensibility. In the preceding view, the juxtaposition of textures presents an agreeable variety of visual 'contrasts', a central concept in Pennant's use of the picturesque: '[O]f Iona, its clustered town, the great ruins, and the fertility of the ground, were fine contrasts, in our passage to the red granite rocks of the barren Mull.'[52] The subsequent description outlined, in an empirical voice, the general shape or topography of the island:

> They [the columns] decreased in height in proportion as they advanced along that face of Staffa opposed to us, or the Eastern side; at length appeared lost in the formless strata; and the rest of the island that appeared to us was formed of slopes to the water edge, or of rude but not lofty precipices.[53]

That the natural colonnade maintained proportional size, as observed at a distance by Pennant, alludes to the role of proportion in the design of columns in architecture.[54] In accounts by both Banks and Pennant architectural analogy and picturesque vision were employed to capture the 'presence' of the landscape for the reader.

The legacy of Banks's picturesque and empirical description is evident in later geological publications such as John MacCulloch's *A description of the western islands of Scotland* (1819). A stickler for empirical accuracy, MacCulloch admonishes, '[t]he language of wonder has already been exhausted on [Staffa]'.[55] Debunking romanticized sentiment of its regular columns that rivalled Grecian examples, MacCulloch remarks that 'its dimensions appear to have been over-rated, in consequence of the mode of measurement adopted, and that the drawings of it which have been engraved, give it an aspect of geometrical regularity which it is far from possessing'.[56] Even so, MacCulloch's attention to accuracy succumbs to a picturesque vocabulary in his descriptions of Fingal's Cave:

> The columns are frequently broken and irregularly grouped, so as to catch a variety of direct and reflected tints mixed with unexpected shadows, that produce a picturesque effect which no regularity could have given.[57]

The accompanying illustration also supports a picturesque vision of the cave entrance and its causeway, as a harmonious composition of light and texture effects (Fig. 9.2). The rough surface of broken causeway columns leads the eye to the standing pillars of the cave. The brightness of the basaltic range is contrasted with the darkened space of the cave's interior. The horizontal lines of the calm ocean waters echo the verticality of the regular columns. The highest stratum of the island blends with the texture of the overcast sky. All lines of the composition appear to radiate out of the darkened aperture. A ship on the left horizon and two staff figures walking up the causeway visually ground the composition. While there is no direct evidence that MacCulloch travelled with a camera obscura, the foreground of the image 'tips' off the page in a fashion similar to scenes traced from this drawing device. Overall, the balance of contrast and the empirical nature of linear styles of drawing suggest a picturesque attention to detail.

While MacCulloch depopulated Staffa's landscape to enhance empirical accuracy of his observations, Banks and James Miller depicted the entire landing party actively engaged in observation of Staffa's physical features.[58]

Figure 9.2 John MacCulloch, *Entrance of Fingal's Cave, Staffa*, from *A description of the western islands of Scotland, including the Isle of Man* (London and Edinburgh, 1819).

Three groups of figures are represented in Miller's pen-and-ink drawing. The first is found at the threshold of Fingal's Cave. In a dinghy, three men paddle about with expressions of admiration on their faces while two figures watch from the water's edge (Fig. 9.1). Two-thirds of the way into the cave, the second group of men observe its interior. The figures appear to be in discussion as one stands and points to a nearby wall while the other peers over the side of the boat to examine the water below. The last group is located to the left of the cave entrance. Here, two individuals inspect the landscape. The figure on the left measures the pillar formation upon which the second figure sits. Surprisingly, the seated person appears to be female. When Banks arrived at Staffa, it was occupied by a family with children.[59] The inclusion of this female figure suggests that one of Staffa's inhabitants accompanied the expedition party during their fieldwork; however, she points to the individual measuring the basalt at her feet, which could imply that her inclusion in the drawing was allegorical in nature.

All of the figures in Miller's drawing appear to be actively engaged with the landscape. They are looking and inspecting, pointing and discussing, and lastly measuring. They appear to be engaged with the specifics of the site in the same way that geologists would later uphold as proper and adequate observational practice. Observation was not a passive activity, and Miller's drawing depicted a landscape covered with figures interacting with their surroundings. Importantly, the naturalists are not represented as overwhelmed or lost by the size of the cavern represented. The height of the figures fills the space of the illustration and gives the naturalist a distinctive presence in the landscape. Later engravings of Fingal's Cave and the Island of Staffa, including the engravings copied from Miller's originals for publication in *A Tour in Scotland*, often reduced the size of the figures and increased the height of the pillars. The image of Fingal's Cave from Faujas de Saint-Fond's *Voyage en Angleterre* (1797), for instance, exaggerated the proportions of the figure in relation to the site (Fig. 9.3).[60] A sketch by James Skene (1775–1864), who accompanied George Bellas Greenough on his Scottish geological tour in 1805, depicted a small group of men dwarfed by a towering island with foaming surge crashing against the causeway (Fig. 9.4).

The engagement of Banks's expedition party with the physical features of Staffa softened expressions of amazement or sublime sensibility in the account. Rather than drawing attention to the emotional narratives of the explorers, the formal properties of the landscape emphasized the object of study: the prismatic effect of the basalt. In a systematic survey of the island, Banks noted and compared variations in the pillars. Beginning at Landing Place, he moved westward towards Cormorant's Cave, after which the pillars diminished along

Figure 9.3 Barthélemy Faujas de Saint-Fond, *Vue de la Grotte de Fingal*, from *Voyage en Angleterre* (1797).

Figure 9.4 James Skene, *Staffa* (1805), chalk on paper.

the western and northern sides of the island.[61] Banks's description abbreviated the un-pillared sections accordingly. He observed that after passing Cormorant's Cave:

> the pillars totally cease; the rock is of a dark brown stone, and no signs of regularity occur till you have passed round the S. E. end of the island (a space almost as large as at that occupied by the pillars) which you meet again on the West side, beginning to form themselves irregularly, as if the stratum had an inclination to that form, and soon arrive at the bending pillars where I began.[62]

Measurement and empirical observation defined the prismatic basalt at the south end of Staffa; however, analogies between natural models and human architectural achievement effectively cushioned an empirical sensibility within an aesthetic discourse that sought to quantify art and taste. In particular, architecture operated as a useful vocabulary for quantification because the study of design depended on measurement and the careful delineation of detailed forms.

While the transition from the initial aesthetic description to the later empirical account appears surprisingly terse, careful reading of Banks's picturesque attention to the ruin/landscape suggests that the shift to empirical observation was not as abrupt as perhaps assumed. The reflections on art and Ossianic history are disrupted with Banks's commanding voice: 'Enough of the beauties of Staffa, I shall now proceed to describe it and its productions more philosophically.'[63] The subsequent 'philosophically' inclined description accurately documented the prismatic basalts as a pseudo-architecture or archetype. Like his fellow dilettanti, Banks meticulously measured and recorded Staffa's ruined form. While reconstruction and preservation of the past were activities allocated to the antiquarian, Banks and his expedition party investigated and recorded the basaltic formations as they reflected on its imagined cultural meaning. Contemporary interest in Gothic and Classical architectures presented a familiar framework for Banks to launch his account of this curious island in which the overall landscape is composed of a set of particulars: the first part being the unusual basaltic pillars, and second being the pediment-like structure of the overhanging tufa.[64] The second half of his account measures and delineates those features in detail.[65]

In essence, Banks's description of Staffa operated like the dilettanti pattern books of Stuart and Revett, and provided a framework for reinventing the past. Picturesque sensibilities directly engaged the appropriation and manipulation of the conceptual content and visualization of objects. The fluid nature of cultural sources and the meanings derived from these ancient arts go hand-in-hand with the ambiguity that the picturesque creates. It disassembled without marking the landscape, and effectively obscured the junctions that might disrupt a composite picture. The picturesque enabled intervention and engagement with Nature for both the naturalist and the local inhabitant. The seated female figure in Miller's pen-and-ink drawing improved her familial landscape though its measurement and accurate delineation, a gesture that would have pleased Pennant who delighted in the improvements he observed on his second tour through Scotland.[66] Banks had spent the day before travelling to Staffa observing the local kelp harvest and industry.[67] As peasants and labourers became extensions of Nature, so too did the naturalist and his habits of witnessing and scientific performance. His inquiry was not merely defined by picturesque expression, but was influenced by an aesthetic behaviour that fundamentally shaped the types of questions asked about Staffa. Primed with an aesthetic vocabulary, Banks and Pennant acted accordingly. Naturalists such as Banks and Pennant were more than just active participants in the process of 'witnessing' natural history, they were agents of aesthetic sensibility who justified and perpetuated a framework for making knowledge and knowing Nature.

Notes

* The author would like to thank Jim Secord, Nick Jardine, Ralph O'Connor, Simon Schaffer and Rory Du Plessis for their encouragement and constructive comments, as well as the Max Planck Institute for the History of Science for supporting this research.

1 Thomas Pennant, *A Tour in Scotland, and Voyage to the Hebrides, MDCCLXXII* 2 vols (Chester, 1774–76).

2 Geoffrey Grigson, 'Fingal's Cave', *Architectural Review* 104 (1948): 51–54, briefly outlines the organic 'origins' of the Gothic style and the integration of Gothic into Classical styles of architecture; Martin Rudwick, 'The Emergence of a Visual Language for Geological Science 1760–1840', *History of Science* 14 (1976): 149–95 (159), suggests that distortions in pictorial depictions of Fingal's Cave 'possibly evoked the neo-classical style'; Barbara Maria Stafford, *Voyage into Substance: Art, Science, Nature, and the Illustrated Travel Account, 1760–1840* (Cambridge, MA and London: The MIT Press, 1984), 59–183, explores the boundaries between natural and artefactual landscapes and architectures; Charlotte Klonk, *Science and the Perception of Nature: British Landscape Art in the Late Eighteenth and Early Nineteenth Centuries* (New Haven and London: Yale University Press, 1996), 74–94, traces the iconography of Fingal's Cave and the sublime affects of Staffa's geological features as depicted by artists.

3 Banks to Thomas Falconer, 12 January 1773, *The Letters of Sir Joseph Banks: A Selection, 1768–1820*, ed. Neil Chambers (London: Imperial College Press, 2000), 35.

4 A. Agnarsdottir, 'Sir Joseph Banks and the Exploration of Iceland', in *Sir Joseph Banks: A Global Perspective*, ed. R. E. R. Banks et al. (Kew: The Royal Botanic Gardens, 1994), 31–48.

5 It was only after inheriting the Overton estate in 1793 that Banks began to privately show interest in studies of the earth. Overton is located in Derbyshire, a region of significant mining heritage and of great interest to students of the earth sciences; see Hugh S. Torrens, 'Patronage and Problems: Banks and the Earth Sciences', in *Sir Joseph Banks*, ed. Banks et al., 49–75.

6 Rhoda Rappaport, 'Borrowed Words: Problems of Vocabulary in Eighteenth-Century Geology', *British Journal for the History of Science* 15 (1982): 27–44; eadem, *When Geologists were Historians, 1665–1750* (Ithaca and London: Cornell University Press, 1997); Martin Rudwick, *The Meaning of Fossils: Episodes in the History of Palaeontology* (Chicago and London: The University of Chicago Press, 1985); idem, *Bursting the Limits of Time: The Reconstruction of Geohistory in the Age of Revolution* (Chicago and London: The University of Chicago Press, 2005), 183–94; Stafford, *Voyage into Substance*; Hugh S. Torrens, 'Geology and the Natural Sciences: Some Contributions to Archaeology in Britain 1780–1850', in *The Study of the Past in the Victorian Age*, ed. Vanessa Brand (Oxford: Oxbow Monograph 73, 1998), 35–59; Noah Heringman, *Sciences of Antiquity: Romantic Antiquarianism, Natural History, and Knowledge Work* (Oxford: Oxford University Press, 2013).

7 Thomas Pennant, *The Literary Life of the Late Thomas Pennant, Esq., by Himself* (London, 1793), 3.

8 Andrew Kennedy, 'Antiquity and Improvement in the National Landscape: The Bucks' Views of Antiquities', in *Tracing Architecture: The Aesthetics of Antiquarianism*, ed. Dana Arnold and Stephen Bending (Malden, MA: Blackwell Publishing, 2003), 68–79, presents a similar argument regarding the folios by Samuel and Nathaniel Buck published between 1726 and 1742.

9 John Bonehill, '"New scenes drawn by the pencil of Truth": Joseph Banks's Northern Voyage', *Journal of Historical Geography* 43 (2014): 9–27.

10 John Gascoigne, *Joseph Banks and the English Enlightenment: Useful Knowledge and Polite Culture* (Cambridge: Cambridge University Press, 1994), 127.

11 Agnarsdottir, 'Sir Joseph Banks', 31.

12 Iona was seen as the original entry point of Christianity from Ireland into Scotland, England and Wales. For history of visitors, see Richard Sharpe, 'Iona in 1771: Gaelic Tradition and Visitors' Experience', *The Innes Review* 63 (2012): 161–259.

13 Roy A. Rauschenberg, 'The Journals of Joseph Banks's Voyage up Great Britain's West Coast to Iceland and to the Orkney Isles, July to October, 1772', *Proceedings of the American Philosophical Society* 117 (1973): 186–226 (206).

14 Rudwick, *Bursting the Limits*, 62–63, 106–8.

15 Martyn Anglesea and John Preston, '"A philosophical landscape": Susanna Drury and the Giant's Causeway', *Art History* 3 (1980): 252–73, 262–63. In 1740 Susanna Drury produced a set of four drawings of the Causeway and won the premium offered by the Dublin Society, an organization that, like the Royal Society of Arts in London, promoted native abilities within agriculture, industry and commerce. For a general survey of the history of columnar basalt, see William B. Ashworth Jr., *Vulcan's Forge and Fingal's Cave: Volcanoes, Basalt, and the Discovery of Geological Time* (Kansas City: Linda Hall Library of Science, Engineering & Technology, 2004). For more on Drury, see Bettie Higgs and Patrick N. Wyse Jackson, 'The Role of Women in the History of Geological Studies in Ireland', in *The Role of Women in the History of Geology*, ed. C. V. Burek and B. Higgs (London: Geological Society Special Publications, 2007), 137–53, 138–39.

16 Banks to Thomas Falconer, 12 January 1773, *Letters of Sir Joseph Banks*, ed. Chambers, 33.

17 Ibid.

18 Pennant, *A Tour in Scotland [...] 1772*, I: 299–300. The original diary entry is almost verbatim with slight changes to punctuation; see Rauschenberg, 'Journals of Joseph Banks': 206.
19 In a similar fashion, the Dissolution of the Monasteries significantly boosted antiquarian interest, as individuals sought out evidence, such as heraldry, to establish their family lineage and right to purchase lands formerly belonging to the Church; see Philip Lindley, *Tomb Destruction and Scholarship: Medieval Monuments in Early Modern England* (Donington: Shaun Tyas, 2007). For the role of the Society of Antiquaries of London and the promotion of popular antiquarianism, see Stephen Bending, 'Every Man is Naturally an Antiquarian: Francis Grose and Polite Antiquities', in *Tracing Architecture*, ed. Arnold and Bending, 100–110.
20 In England, the reception of the Ossianic epics as collected by MacPherson was mixed; see Colin Kidd, *Subverting Scotland's Past: Scottish Whig Historians and the Creation of an Anglo-British Identity, 1689–1830* (Cambridge: Cambridge University Press, 1993). On reception of MacPherson's account in Europe, see Howard Gaskill, ed., *The Reception of Ossian in Europe* (London: Bloomsbury Academic, 2004).
21 MacPherson was sponsored by members of Edinburgh's elite, the Poker Club, to recover the Ossianic epics. See Stafford, *Voyage into Substance*, 115–16, 153. Individuals who expressed interest and supported the project included: Hugh Blair, Alexander Carlyle, David Dalrymple, Adam Ferguson, John Home and David Hume. Blair was a keen supporter of MacPherson's translation. For a detailed study on MacPherson's methodology, see Kidd, *Subverting Scotland's Past*, 228–39. On Blair's theories on the natural history of language, see Matthew Eddy, 'The Line of Reason: Hugh Blair, Spatiality and the Progressive Structure of Language', *Notes and Records of the Royal Society* 65 (2011): 9–24.
22 MacPherson explicitly draws parallels between Ossian and Pope's translation of Homer in his footnotes; see James MacPherson, *Fingal, an ancient epic poem, in six books: together with several other poems, composed by Ossian the son of Fingal* (London, 1762), 12–13, 18, 85, 177–78, 238, 265, 298.
23 Rauschenberg, 'Journals of Joseph Banks': 205.
24 Pennant, *Tour in Scotland [...] 1772*, I: 302.
25 Arnaldo Momigliano, 'Ancient History and the Antiquarian', *Journal of the Warburg and Courtauld Institutes* 13 (1950): 285–315. Maria Grazia Lolla, 'Monuments and Texts: Antiquarianism and the Beauty of Antiquity', in *Tracing Architecture*, ed. Arnold and Bending, 11–29, examines the importance of objects for both dilettanti and antiquarian groups. The dilettanti and art theorist Johann Joachim Winckelmann (1717–1768) wanted to render monuments such as statues, marble and terra cotta bas-reliefs, gems and paintings as indistinguishable from text. The Society of Antiquaries of London sought to produce a facsimile of the Doomsday Book, which focused their attention on the materiality of the copy. In both instances, objects were used to 'correct' text.
26 See Jason M. Kelly, *The Society of Dilettanti: Archaeology and Identity in the British Enlightenment* (New Haven and London: Yale University Press, 2009), 11–19. A full spectrum of political meaning was attributed to 'Gothic' during the eighteenth century. In the 1730s, associations between Gothic architecture and the English constitution came to be a patriotic symbol; see Alexandrina Buchanan, 'Interpretations of Medieval Architecture, *c.*1550–*c.*1750', in *Gothic Architecture and its Meanings 1550–1830*, ed. Michael Hall (Reading: Spire Books, 2002), 27–50, 43. David Stewart, 'Political Ruins: Gothic Sham Ruins and the '45', *Journal of the Society of Architectural Historians* 55 (1996): 400–11,

argues that in some instances Gothic ruins were erected as political demonstrations call-
ing for a break from the past. For an overview on the political meanings of 'Gothic', see
Michael Hall, 'Introduction', in *Gothic Architecture*, ed. idem, 7–24, 18–24.

27 Mordaunt Crook, *The Greek Revival: Neo-Classical Attitudes in British Architecture, 1760–1870*
(London: John Murray, 1972), 16, quoting Stuart and Revett from *Antiquities of Athens*.

28 On Staffa as a site that combines Classical and Gothic architectural elements, see
Grigson, 'Fingal's Cave'.

29 William Gilpin, *An essay upon prints; containing remarks upon the principles of picturesque beauty*
(London, 1768), 1–2.

30 Gilpin, *Essay upon prints*, 2.

31 Pennant, *Tour in Scotland [...] 1772*, I: 300–301.

32 Banks measured the tallest point on Staffa at 128 feet; the geologist John MacCulloch
later measured it at 112 feet.

33 Pennant, *Tour in Scotland [...] 1772*, I: 301.

34 The Society of Dilettanti spent nearly £30,000 sponsoring expeditions to Greece and
publishing discoveries made on those trips between 1734 and 1852, see Crook, *Greek
Revival*, 62.

35 See Eileen Harris and Nicolas Savage, *British Architectural Books and Writers, 1556–1785*
(Cambridge: Cambridge University Press, 1991), 439–50, for an account of their tour.
On British travellers in Greece in the early nineteenth century, see Helen Angelomatis-
Tsougarakis, *The Eve of the Greek Revival: British Travellers' Perceptions of Early Nineteenth-
Century Greece* (London: Routledge, 1990). On subsequent 'expansions' of the Grand
Tour itinerary, see Brian Dolan, *Exploring European Frontiers: British Travellers in the Age of
Enlightenment* (London: MacMillan, 2000).

36 *Edifices de Rome* functioned as the standard for depicting ancient architecture until
Antiquities of Athens was published; see David Watkin, *The English Vision: The Picturesque in
Architecture, Landscape and Garden Design* (London: John Murray, 1982), 26. On *Antiquities
of Athens* as a pattern book, see Crook, *Greek Revival*, 17; Harris and Savage, *British
Architectural Books*, 431–37. On the variety of pattern books available during the eighteenth century, see ibid.; Michael McCarthy, *The Origins of the Gothic Revival* (New Haven
and London: Yale University Press, 1987), 4–26.

37 Bruce Redford, *Dilettanti: The Antic and the Antique in Eighteenth-Century England* (Los
Angeles: J. Paul Getty Museum and Getty Research Institute, 2008), 46–49; Frank
Salmon, *Building on Ruins: The Rediscovery of Rome and English Architecture* (Aldershot:
Ashgate, 2000), 41. On eighteenth-century criticism of *Edifices de Rome*, see Salmon,
Building on Ruins, 35–39.

38 Salmon, *Building on Ruins*, 19.

39 Dana Arnold, 'Facts or Fragments? Visual Histories in the Age of Mechanical
Reproduction', in *Tracing Architecture*, ed. Arnold and Bending, 30–48, 40–42; Salmon,
Building on Ruins, 46–47. John Soane's installation of architectural fragments at Lincoln's
Inn further exaggerated the abstraction of elements.

40 William L. MacDonald, *The Pantheon: Design, Meaning, and Progeny* (London: Allen
Lane, 1976), 110–11, describes the dome as a circular canopy signifying the heavens. It is as wide as it is broad, meaning it is perfectly circular. The Pantheon has
historically been represented as an isolated structure; however, it was seventeenth- and
eighteenth-century building programmes that eventually rendered the Pantheon a
free-standing building. In a period of destabilized artistic and architectural canons,
Piranesi attempted to reconstruct the Pantheon as part of a larger Roman complex; see

Susan Dixon, 'Piranesi's Pantheon', in *Architecture and Experience: Radical Change in Spatial Practice*, ed. Dana Arnold and Andrew Ballantyne (London and New York: Routledge, 2004), 57–80.

41 Christopher Hussey, *The Picturesque: Studies in a Point of View* (London and New York: G. P. Putnam's Sons, 1927), argues that the picturesque is not a style, but a mode of combining and using styles. For an overview of picturesque landscape design, see Mavis Batey, 'The Picturesque', *Garden History* 22 (1994), 121–32. On politics of the picturesque, see John Barrell, *The Dark Side of the Landscape: The Rural Poor in English Painting, 1730–1840* (Cambridge: Cambridge University Press, 1983). On the picturesque in travel literature, see Nigel Leask, *Curiosity and the Aesthetics of Travel Writing, 1770–1840: 'from an antique land'* (Oxford: Oxford University Press, 2002).

42 Gilpin, *Essay upon prints*, 30.

43 Ibid. 31.

44 Dixon, 'Piranesi's Pantheon', 58, uses the term 'multi-informational' to describe the picturesque work of Jean-Pierre Hoüel (1735–1813).

45 Stuart and Revett accused Julien David Le Roy of picturesque distortions in his *Le ruines de plus beaux monuments de la Grèce* (1758). In Le Roy's work, atmospheric qualities were favored over accurate dimensions of buildings for a romantic vision of Greece; see Redford, *Dilettanti*, 53–59. Robert Adams was criticized for the same fault; see ibid. 81–82. On the picturesque nature of Stuart's and Revett's images, see David Watkin, 'Stuart and Revett: The Myth of Greece and its Afterlife', in *James 'Athenian' Stuart, 1713–1788: The Rediscovery of Antiquity*, ed. Susan Weber Soros (New Haven and London: Yale University Press, 2007), 19–57 (26–27).

46 Grigson, 'Fingal's Cave': 54. See also John Summerson, *Georgian London* (1945; New Haven and London: Yale University Press, 2003), 232–35.

47 Uno von Troil, *Letters on Iceland: containing observations on the civil, literary, ecclesiastical, and natural history; antiquities, volcanoes, basalts, hot springs; customs, dress, manners of the inhabitants, &c. &c. made, during a voyage undertaken in the 1772, by Joseph Banks, Esq.* (Dublin, 1780), 273–74. The naturalist Saussure suggested the Louvre, St Peter's and Palmyra were architectural examples that Troil imagined for comparison; see L. A. Necker de Saussure, *A voyage to the Hebrides, or western islands of Scotland; with observations on the manners and customs of the highlanders* (London: Sir Richard Phillips & Co., 1822), 31. Debates regarding true 'Classical' design in the history of architecture is a long and fascinating history beginning with the controversial columns that Charles Le Brun (1619–1690), Louis Le Vau (1612–1670) and Claude Perrault (1613–1688) proposed for the Louvre; see Harry Francis Mallgrave, *Modern Architectural Theory: A Historical Survey, 1673–1968* (Cambridge: Cambridge University Press, 2007), 39. On the continuation of the Graeco–Roman controversy during the eighteenth century, see John Wilton-Ely, *Piranesi as Architect and Designer* (New Haven and London: Yale University Press, 1993), 35–41. The dome of the new St Peter's Basilica, designed by Bramante in 1506, was intended to surpass, in scale and technology, all the greatest monuments in Rome including the Pantheon; see Charles B. McClendon, 'The History of St Peter's Basilica, Rome', *Perspecta* 25 (1989), 32–65 (48); MacDonald, *Pantheon*, 112. In 1753 Robert Wood published *The ruins of Palmyra*. Like *Antiquities of Athens*, it was conceived in association with Desgodetz's treatise. The publication was in part sponsored by the Society of Dilettanti and continued to demonstrate the Society's position that architectural innovation existed outside of Rome, see Harris and Savage, *British Architectural Books*, 491–92; Salmon, *Building on Ruins*, 41.

48 On the difficulties of defining the meaning of 'picturesque', see Stephen Copley and
    Peter Garside, 'Introduction', in *The Politics of the Picturesque: Literature, Landscape and
    Aesthetics since 1770*, ed. eidem (Cambridge: Cambridge University Press, 1994).

49 Pennant, *Tour in Scotland [...] 1772*, I: 300.

50 Ibid.

51 Ibid. 298.

52 Ibid.

53 Ibid.

54 In architectural theory, proportion determined the classification and utility of a column.
    For an overview of the history of the column and debates surrounding proportion, see
    Hanno-Walter Kruft, *A History of Architectural Theory: From Vitruvius to the Present*, trans.
    Ronald Taylor, Elsie Callandar and Antony Wood (New York and London: Zwemmer
    and Princeton Architectural Press, 1994).

55 John MacCulloch, 'On Staffa', *Transactions of the Geological Society of London* 2 (1814),
    501–9 (506).

56 Ibid.

57 John MacCulloch, *A description of the western islands of Scotland, including the isle of Man*,
    3 vols. (London: A. Constable, 1819), II: 17.

58 On the politics of emptying or improving picturesque landscape, see Ann Bermingham,
    Landscape and Ideology: The English Rustic Tradition, 1740–1860 (London: Thames &
    Hudson, 1987); Barrell, *Dark Side of the Landscape*; idem, 'The Public Prospect and
    the Private View: The Politics of Taste in Eighteenth-Century Britain', in *Reading
    Landscape: Country–City–Capital*, ed. Simon Pugh (Manchester: Manchester University
    Press, 1990), 19–40; John Macarthur, 'The Butcher's Show: Disgust in Picturesque
    Aesthetic and Architecture', *Assemblage* 30 (1996): 32–43; Sam Smiles, 'Dressed to Till:
    Representational Strategies in the Depiction of Rural Labour, *c.*1790–1830', in *Prospects
    for the Nation: Recent Essays in British Landscape, 1750–1880*, ed. Michael Rosenthal,
    Christina Payne and Scott Wilcox (New Haven and London: Yale University Press,
    1997), 79–95.

59 Rauschenberg, 'Journals of Joseph Banks', 206. The expedition tent was small for a
    party of nine men. Five men ended up sleeping in the local inhabitants' house.

60 See Rudwick, 'Emergence of a Visual Language': 173–74, Klonk, *Science and the Perception
    of Nature*, 81–82; Donald B. MacCulloch, *Staffa* (1927; Newton Abbot, London and
    North Pomfret, VT: David and Charles, 1975), 20–22.

61 Cormorant's Cave is labelled 'McKinnons Cave' on John MacCulloch's map from
    Description of the western islands.

62 Pennant, *Tour in Scotland [...] 1772*, I: 308.

63 Ibid. 302.

64 Momigliano, 'Ancient History', 306.

65 On the use of accuracy to establish knowledge in antiquarian visual representation, see
    Sam Smiles, *Eye Witness: Artists and Visual Documentation in Britain, 1770–1830* (Aldershot
    and Brookfield: Ashgate, 2000).

66 Pennant, *Literary Life*, 15. On Constable and representations of the countryside as an
    extension of his person, see Bermingham, *Landscape and Ideology*, chapter 3.

67 Rauschenberg, 'Journals of Joseph Banks', 205–6.

Chapter 10

PENNANT, HUNTER, STUBBS AND THE PURSUIT OF NATURE

Helen McCormack

And what knowledge can be more useful than of those objects with which we are most intimately connected?

– Thomas Pennant, *British Zoology* (1776–77)

Thomas Pennant, William Hunter and George Stubbs shared a remarkably similar ambition for the ways in which the fine arts might be put to use in the production of knowledge of natural history in the second half of the eighteenth century. Each understood how the activities of naturalists encompassed Enlightenment values by considering the study of the natural world in 'an enlarged view'.[1] In this sense, and for Stubbs and Hunter in particular, this meant incorporating anatomy within the broader realm of the natural sciences that typically featured subject matters such as geology and zoology, alongside astronomy and knowledge of climates. During the 1770s Thomas Pennant's close observational approach to the natural world became explicitly connected to the work of William Hunter and George Stubbs and consequently featured among the debates surrounding the imitation and representation of nature within the cosmopolitan world of the Royal Academy of Arts and polite culture in London.[2] How Pennant's research for his publications of natural history, *British Zoology* and *Arctic Zoology*, and for his tours of Wales and Scotland contributed to an 'empirical habit of vision', in common with Hunter and Stubbs, is described in this essay.[3] Pennant also understood that the relationship between travel and natural history was inextricable; alongside the pursuit of antiquities, natural history involved traversing the landscape, locally and nationally, in order to carry out field studies. Travel was an accepted preoccupation of the naturalist and antiquarian: '& you well know that natural history is not to be acquired in a closet', as the antiquarian Thomas Falconer remarked to Sir Joseph Banks, alluding to the conventional view of the savant immersed in his private study, closed off from the elements.[4] While Pennant was a gentleman with time to spare on his 'disinterested' pursuit of natural knowledge, William Hunter and George Stubbs were both professional men who earned their living as anatomist and artist respectively. Their work kept them confined to the city but their regular correspondents in the interconnected communities of art and science provided them with the types of intercourse necessary to maintain national and European-wide Enlightenment objectives.[5]

All three men, very close in age (Thomas Pennant, 1726–1798; William Hunter, 1718–1783; George Stubbs, 1724–1806), contributed to an analysis of the ways in which knowledge of the animal world could be represented by means of the fine arts: zoology, 'the noblest part of natural history', informed an understanding of the interconnections between all living, sensitive beings.[6] Among their published writings, correspondence and illustrations are many references to a strongly held sense of empirical method, originating in the writings of John Locke in large part, but in other aspects deriving from an anatomical tradition of *autopsia*, or 'to see with one's own eyes'.[7]

Visual culture of the Enlightenment era depicted representation as a form of knowledge, as such, anatomy, as with other branches of the natural sciences, was closely interconnected with the fine arts more generally and William Hunter's lectures to students at the Royal Academy of Arts made this connection powerfully explicit. However, anatomical knowledge was also often characterized as a form of metaphor, applied to a range of ideas and associations during the period. Anatomy lent itself naturally to caricature and satire especially (anthropomorphized maps of the British Isles and images of the 'Body Politic' are just some examples). Barbara Maria Stafford has explained that the body 'As a visible natural whole made up of invisible dissimilar parts [...] was the organic paradigm or architectonic standard for all complex unions.'[8] Despite the compelling visual appeal of anatomy as metaphor, William Hunter's teaching suggests 'a completely unmediated vision' eschewing any 'misrepresentation' of the body and its sub-surface.[9] Landscape artists in particular may have been attracted to an image of the land as body: 'Spurred by meteorological investigations [...] This physical but subtle reality seemed to lie beyond the reach of ordinary perception', incorporating a networked subcutaneous membrane that appeared to provide the topographically visible (and contested) landscape with a unifying purpose.[10] Arguably, however, such metaphoric language did not fulfil the objectives of accuracy and immediacy required by naturalist empiricists such as Pennant, Hunter and Stubbs.

These ideas appear to be confirmed by the emergence of what is often referred to as a 'truth to nature' demanded by such an empirical process.

Pennant, Hunter and Stubbs all claim that to represent the natural world 'truthfully' means copying directly from nature and not from facsimiles. For Pennant this meant undertaking extensive travels and correspondence, for Hunter and Stubbs it involved forensic examination of the human and animal body.

Recent writings on the entanglements of meanings attached to such phrases as 'truth to nature' or 'empiricism' in the period have concluded that these were hardly fixed or clearly defined features of artworks.[11] Often those images that claimed to be the most 'truthful' or 'objective' were distinctly neither, and artists submitted to a range of subjective experiences infiltrating their 'scientific' procedure. Such intricacies are described here in a comparison between the writings of the three men, in the images produced alongside their researches, and in the ways that each thought an appropriate means of representing and imitating nature.

Animalia

Thomas Pennant's *Literary Life* (1793) records how 'A present of the ornithology of Francis Willughby, esq. made to me, when I was about the age of twelve, first gave me a taste for that study, and incidentally a love for that of natural history in general, which I have since pursued with my constitutional ardor.'[12] After initially publishing 'An Account of Some Fungitae and Other Curious Coralloid Fossil Bodies' in *Philosophical Transactions* (1755), Pennant resolved to begin work on *British Zoology* in 1761, with the first volume published in 1766. The preface to this book includes some striking commentary on the study of natural history in Britain and on its usefulness to various industries, such as lead ore, copper and silver mines: 'The Haematites iron ores of Cumberland, and the beautiful columnar irone ores of the Forest of Dean, are sufficient to display our riches in that useful commodity.'[13] More interestingly, Pennant uses the preface as a discourse on the relationship between natural knowledge and the polite arts, 'which have hitherto been too little connected with it.'[14] Pennant explains that knowledge of the natural world provides practical benefits for artists, in knowing the origins of pigments:

> To instance particularly painting, its uses are very extensive: the permanency of colours depends on the goodness of the pigments; but the various animal, vegetable, and fossil substances [...] can only be known by repeated trials; yet the greatest artists have failed in this respect: the shadows of the divine Raphael have acquired an uniform blackness, which obscures the finest productions of his pencil, while the paintings of Holbein, Durer, and the Venetian School still exist in their primitive freshness.[15]

The materiality that makes up the very paint used by artists has a direct connection to the natural world for Pennant, but the rhetorical aspects of painting and its relationship to poetry, especially, are scrutinized by his naturalist's eye. The benefits of knowledge of such materials however are small, says Pennant, when compared to the advantages of an extensive understanding of natural forms: 'Painting is an imitation of nature; now who can imitate without consulting the original?'[16] The persuasiveness of a painting's depictions of animals, vegetables and landscapes is diminished if these are not experienced by the artist himself or herself; and Pennant suggests this is why such objects are usually treated as secondary to the overall subject of a painting, strewn around the outer edges of an historical scene or portrait and lacking in any 'correctness of design'. It is in this passage that Pennant appears to be addressing the very purpose of academic forms of painting becoming fashionable in late eighteenth-century Britain.[17]

As with his extensive *Tours*, Pennant makes space in *British Zoology* for comment on broader cultural and political matters. In pursuing knowledge of natural history, the 'country gentleman' would be 'laying up a fund of useful knowledge; they would find their ideas sensibly enlarged'.[18] That these enlarged ideas would comprise the minutiae of descriptions of the smallest living beings to the survey of vast tracts of improved or wild countryside is the attraction of Pennant's schemes; as passages from *British Zoology* intersect with reflections on his tours around Britain, his readers in the eighteenth century could make direct connections between the centres of contemporary cosmopolitan culture and the periphery, the geographical limits of *Ultima Thule*, albeit in relatively positivist terms.[19] In fact, *British Zoology* appears in print at a particular moment in the development of the fine arts in Britain, one that is often dominated by the role of the Royal Academy of Arts and its pedagogical aims to inculcate a specific civic humanist agenda. However, *The Triumph of the Arts*, a poem written to commemorate its foundation in 1768, reflects the Academy's original aims:

> The Sun of Science beams a purer ray;
> Behold! A brighter train of Year,
> A new Augustan Age appears […]
> Where Art may join with Nature and with Sense
> Splendour with Grace, with Taste Magnificence.[20]

The emphasis on arts, science and nature in these celebratory lines indicate the purpose of the Academy, where students and academicians would lead the promotion and development of the sciences, or useful knowledge, in the period. In particular, Dr William Hunter regularly read lectures to students on the anatomy of the human body.[21]

By the 1770s William Hunter was, like Pennant, engaged in several personal projects, the most important being research for his impressive *The Anatomy of the Human Gravid Uterus* (1774). During this time, Hunter had entered into correspondence with Thomas Pennant regarding the latter's *Synopsis of Quadrupeds* (1771). In this book Pennant attempts to bring together a more systematic approach to the definitions of each animal so far collected in the work of George-Louis Leclerc, Comte de Buffon (1707–88) and Carl Linnaeus (1707–78). His preface acknowledges their contributions to the overall publication:

> So far of System: the rest of my plan comprehends numerous Synonyms of each Animal, a brief description, and as full an account of their place, manners, or uses, as could be collected from my own observations, or the information of others; from preceding Writers on the subject; from printed Voyages of the best authorities, or from living Voyagers.[22]

In this sense, Hunter's contribution might also be considered as the authority of a 'living voyager'. Hunter had been observing two Indian antelopes, or nilgai, which had been gifted to Queen Charlotte, and in 1771 he read a paper to the Royal Society presenting his findings. 'An Account of the Nyl-ghau, an Indian Animal not hitherto described' was the culmination of Hunter's researches into the defining characteristics of animal species and the discovery of new types.[23] In 1769 he commissioned George Stubbs to paint *The Nilgai* (Fig. 10.1) in order to have an accurate copy of the animal's likeness. He pointed out in his presentation the advantage of an image over a written description of such unusual animals:

> Good paintings of animals give much clearer ideas than descriptions. Whoever looks at the picture which was done under my eye, by Mr. Stubbs, that excellent painter of animals, can never be at a loss to know the Nyl-ghau, wherever he may happen to meet with it.[24]

This belief in the capacity of an image to inform and instruct in a way not readily reducible to language conforms with Pennant and Hunter's antiquarian approach to the study of natural history.[25] In this example Stubbs is given credit for his ability to follow Hunter's precise instruction, producing a picture 'done under my eye'. However, the nilgai is portrayed in a domesticated landscape and the painting describes more than the physical features of the animal itself. Stubbs depicts something of the animal's temperament, as expressed by Hunter in his anthropomorphized description:

Figure 10.1 George Stubbs, *The Nilgai* (1769), oil on canvas.

All of the time that the two of them were in my stable, I observed this particu-larity, *viz*, that whenever any attempt was made upon them, they immediately fell down upon their fore-knees [...] but, as they never darted, I so little thought this posture meant hostility, that I rather supposed it expressive of a timid or obsequious humility.[26]

This is clearly expressed in Stubbs's painting of the animal and, read along-side Hunter's description, his expressive, anatomically accurate image creates a sensibility that defies narrow interpretations of naturalism such as those attributed to Sir Joshua Reynolds. In a much-quoted reference, Reynolds tells his students that 'a mere copier of nature can never produce anything great; can never raise and enlarge the conceptions, or warm the heart of the spectator'.[27] Remarks such as these appear throughout Reynolds's *Discourses* and have contributed to a rather limited interpretation of certain works pro-duced during the Academy's early years. Reynolds's promotion of a general-ized, or abstracted, rather than a particularized nature was criticized in his own day, perhaps most vehemently by one of Hunter's own draughtsmen,

Jan van Rymsdyk (d. 1790), who undoubtedly targeted the president as one of the 'nature-menders'.[28] In his defence of his own work and in an attack on the style of painting emerging from the Academy, Rymsdyk declared:

> I am obliged to give an answer why this Country has not been so happy in pro-
> ducing such good painters as poets [...] To which we answer, that these Men
> followed Nature; and the Painters not, but were only Nature-Menders.[29]

Inevitably, Reynolds's *Discourses* contributed to a degree of tension surround-ing the appropriate representation and imitation of nature and perhaps Stubbs would have been considered alongside Rymsdyk as an enemy to 'nature-menders, mannerists, or antiques &c'.[30] Nor did Stubbs's own trou-bled relationship with the Academy allow for a more definitive place for his work beyond that of a mere 'animal painter'.[31]

While Reynolds's *Discourses* became the predominant channel for the dis-cussion of the fine arts in Britain during the eighteenth century, there appears to have been a shift in his views, particularly his ideas relating to the deline-ation of the natural world in Dutch and Flemish works, directly before and soon after his visit to Holland in the late summer of 1781. Reynolds's *Discourses* VII and VIII contain complementary references to 'those ornamental parts of our Arts', most likely in relation to the particular minuteness of natural objects in Dutch painting.[32]

The opinions of academicians such as Reynolds and Hunter were influen-tial on the ways that naturalists proceeded in their work inscribing knowledge of Britain's material environment, as the fine arts developed a new 'visual dis-course' that identified archaeology, topography, anatomy, zoology and antiq-uities, alongside the natural productions of the nation.[33] Indeed, anatomical atlases such as Hunter's *The Anatomy of the Human Gravid Uterus* (1774), with its exquisitely engraved life-sized plates, is, perhaps, the exemplar that unites medical illustration with the fine arts to represent natural knowledge in its original reality.[34]

These ideas are palpable in Pennant's *Tours* of Wales and Scotland, as the author strives to achieve a level of description in both narrative and imagery that might match his own profoundly empirical standards. Pennant made two trips to Scotland: in 1769 (published in 1771) and in 1772 (pub-lished in 1774–76). The purpose of the second tour was 'to render more complete my preceding tour'.[35] For a number of reasons this tour was more successful, due mostly to the assistance Pennant gained from notable experts throughout the country.[36] The finished publication also contains many more images, the majority of which were drawn by Moses Griffith (1747–1819). The Scottish *Tours* have sometimes proved controversial and Pennant has been accused of being an interloper, someone who stayed too short a time to make any significant impact on his environment. Equally, he has been considered an enlightened visionary, travelling to North Britain in order to witness the project of improvement and take account of the transformation of Scotland into a commercial country, fully absorbed within the progressive British imperium.[37]

At many stages during his travels, Pennant considers the diverse array of animals he encounters and, like Stubbs's imagery, his descriptions can be highly expressive, providing the reader with a vivid account of the life and natural habitats of British wildlife – descriptions that, when read alongside illustrations, often go beyond practical knowledge. For example, at Lochranza, on the Isle of Arran (Fig. 10.2), Pennant encounters a basking shark, 'harpooned some days before', and he travels across the bay to examine the 'perfect monster' for himself. He then recounts the physiognomy of the creature in his text, including its length, breadth, 'its upper jaw much longer than the lower. The teeth minute, disposed in numbers along the jaws [...] These fish are called in the Erse, *Cairban*; in the Scotch, *Sailfish*, from the appearance of the dorsal fins above water.'[38] Accounting for some features of the animal's habitat and migratory routes as explained by Linnaeus, Pennant proceeds to describe the dramatic nature of the suffering caused by the use of the harpoon in capturing and killing sharks:

Figure 10.2 Moses Griffith, *Loch Ranza Bay. And the manner of taking the basking shark*, etching by P. C. Carnot (1774).

As soon as they perceive themselves wounded, they fling up their tail and plunge headlong to the bottom, and frequently coil the rope around them in their agonies, attempting to disengage themselves from the weapon by rolling on the ground, for it is often found greatly bent.[39]

Despite having arrived at Lochranza some days after the shark had been killed and brought onto the shore, Moses Griffith's illustration to accompany the episode depicts the dramatic scene of the shark's killing. The image shows a calm loch with some fishing boats near the shark and the animal flailing in the water, just as the harpoon is about to be plunged into its flesh. Rather than presenting a clear visual description of the shark, as in Hunter's claim for Stubbs's animal imagery, Griffith has employed the conventions of a landscape sketch, depicting the powerful backdrop of the high 'theatre of mountains' of Arran to create a dark mood surrounding the inevitably brutal action about to take place. Pennant's text helps to evoke tension by carefully crafting such sublime scenes in writing, directing the way the eye of the reader should look over this reimagined event.

In this sense, Pennant's text and Griffith's imagery combine to present the reader with an immediacy far beyond that required by naturalist field recording or systematic classification. In fact, one of the criticisms of narratives of travel and exploration was that literal descriptions were too objective. As mentioned earlier in this essay, Thomas Falconer, in his lengthy correspondence with Sir Joseph Banks, considered the employment of artists on expeditions and voyages as crucial to the advancement of knowledge of natural history. Falconer was particularly critical of Martin Martin's *A Description of the Western Islands of Scotland* (1703), because it relied too heavily on written descriptions:

> For this reason Martin and others have given us a false idea of the Western islands […] What an assistance it is then to truth to have the objects delineated by one common measure which speaks universally to mankind.[40]

As for Pennant, Hunter and Stubbs, drawing and the skills of the draughtsman provided a universal language for Falconer and Banks: a well-drawn image could often convey knowledge in a much clearer and more precise fashion than a written account.

However, it was not only the objective record that artists would incorporate within their imagery: a sense of their subjective experiences also contributed to the 'immediacy and plausibility of the pictorial account'.[41] Pennant's writings often reflect the irrepressible experiences of field study and physical encounters that could not be subdued by recourse to conventions of stark representation. Where he had to rely on images previously provided, however, this sense of emotional experience was usually lost. In a highly charged account of a seal cull in Caithness, the accompanying plate, showing seals taken from a painting in the Museum at Oxford, fails to complement the sensitivity expressed by the author in his text. The vast caves of Caithness are immense openings into the sea and provide a breeding place for seals between October and November. Here Pennant describes a scene of cruel activity:

> The seal hunters enter the mouths of the caverns about midnight, and rowing up as far as they can, they land, each of them being provided with a bludgeon, and properly stationed, light their torches and make a great noise, which brings down the seals from the further end in a confused body with frightful shrieks and cries [...] [they] kill as many as straggle behind (chiefly the young).[42]

Depictions of cruelty and savagery in wild nature – of man's callous treatment of animals, as well as the ferocity of animal behaviour towards one another – were increasingly fascinating topics of fine art productions by the second half of the eighteenth century. The exemplar of such explorations of sublime experience in nature is, of course, George Stubbs's famous series of paintings of a horse being attacked by a lion, which he began in the early 1760s. In *British Zoology* Pennant praises Stubbs's ability to capture the diversity of animal natures:

> Mr. Stubbs, an artist not less happy in representing animals in their still moments, than when agitated by their furious passions; his matchless paintings of horses will be lasting monuments of the one; and that of the lion and panther of the other.[43]

Stubbs's gift for conveying both ferocious passions and still moments in his studies of these animals is no doubt attributable to his intense and extensive knowledge of anatomy and his highly trained skills of *autopsia*. In his memoir of Stubbs, Ozias Humphry remarked that the artist, on being instructed to copy from a series of Old Master paintings, refused to entertain the idea and instead vowed only to work directly from nature.[44] The empirical origins of Stubbs's images single them out from the conventions of both Neoclassicism and Romanticism; like Pennant's descriptions and Hunter's researches, Stubbs's artworks are revelatory, deeply integrated within empirical scientific knowledge of the mid- to late eighteenth century.[45] This is a period characterized by a move away from mechanistic and mathematical theories of living organisms to an intense exploration of the life force of matter itself, a time of experimental anatomy reflected in Stubbs's remarkable productions.[46]

Therefore, in their 'stillness' Stubbs's paintings for William Hunter might appear to be exactly as the anatomist had intended, depicting the animal in all its anatomical accuracy. As with Stubbs's *Nilgai*, however, his painting of *The Moose* (1770) owes more to the sublime meanings of Edmund Burke's essay and the subjective expressiveness of the artist's imagination than might at first seem apparent. It is clear that Pennant selected Stubbs's painting as Plate VIII (Fig. 10.3) in *Arctic Zoology* exactly for its *vita vitalis*, its extreme anatomical tension.[47] Hunter's manuscript notes 'Various Notes on the Orignal – Canadian name for the Moose' reiterate his conviction that a drawing executed directly from nature constitutes a form of universal language:

> Yet for many purposes, especially in the Arts, and in Natural History there is a language which is both easily acquired, & tho' not so copious, is more expressive than any language in the world, and at the same time so plain that the unlearned as well as the learned, understand it at first sight: I mean the art of drawing.[48]

Figure 10.3 George Stubbs, *The Moose* (1770), oil on canvas.

These sentiments are also revealed in Pennant's selection of some lines from James Thomson's poem 'Winter':

> Rais'd o'er the heapy wreath, the branching elk,
> Lies slumbering sullen in the white abyss.
> The ruthless hunter wants nor dogs nor toils;
> Nor with the dread of sounding bows he drives
> The fearful flying race, with ponderous clubs,
> As weak against the mountain heaps they push
> Their beating breast in vain, and piteous bray,
> He lays them quivering on th' ensanguin'd snows,
> And with loud shouts rejoicing bears them home.[49]

While Pennant's description of the moose seems rather insensitive – 'The vast size of the head, the shortness of the neck, and the length of the ears, give the beast a deformed and stupid look'[50] – as with his depiction of the seal cull in Caithness, it is the means of hunting and killing these inoffensive creatures that occasions his sympathies:

> The hunters avoid entering on the chase till the sun is strong enough to melt the frozen crust with which the snow is covered, otherwise the animal can run over the firm surface: they wait till it becomes soft enough to impede the flight of the Moose; which sinks up to the shoulders, flounders, and gets on with great difficulty.[51]

The lines from Thomson's poem, set alongside Stubbs's illustration of the moose, demonstrate the commitment to accuracy and authenticity of Pennant's writings. In his introduction to *Arctic Zoology*, Pennant returns to his own experiences of travel throughout Britain, incorporating Scotland within the vast expanse of the northern hemisphere alongside Greenland, Iceland and the remoter isles, Faroe and Fair Isle, in his 'departure northwards'.[52] Thomson's poetry achieved the ultimate form of naturalistic representation having initiated a blank-verse style that encapsulated *pure* description of the natural world. Thomson's popular and enduring verses established, in a general sense, that as the objects of nature were appropriate subjects for poetry, so they were entirely suitable subjects for painting:

> [I]t was, however, thought that they could not legitimately constitute the whole, or even the principal part of a capital piece. Something of a more solid nature was required as the ground work of a poetical fabric.[53]

In his essay in response to Thomson's *The Seasons*, the surgeon–poet John Aikin argues that Thomson had rejected the 'antient' taste for an 'abstracted' nature:

> The most faithful pencil here produces the noblest pictures; and Thomson by strictly adhering to the character of the poet of nature, has treated all these topics with a true sublimity.[54]

Therefore, exactness and accuracy creates noble imagery, drawn with a faithful or truthful pencil, exhorting a sublime experience in the reader and made even more pronounced, in Pennant's work, by the inclusion of Stubbs's anatomical exactness.

One criticism of Thomson's poetry, for Aikin, is that *The Seasons* neglects to explore the interconnections between man and his environment: 'But the rural landskip is not solely made up of land, and water, and trees, and birds, and beasts; *man* is a distinguished figure in it.'[55] Aikin argues that modern [scientific] philosophy presented the opportunity to explore these affections of physical sensations and intellectual enquiry. In his voyage northwards, Pennant employed a similar objective:

> The manners of the people ought not less to be attended to; and their changes, both mental and corporeal, by comparison of the present state of remote people with nations with whom they had common ancestors, and who may have been discovered still to retain their primaeval seats.[56]

Aikin's essay 'on the Plan and Character' of Thomson's poem might also be understood as a response to the physico-theology represented in the poet's writings in his own time – an aspect of Thomson's poetry that often situates him within an earlier period of Newtonian (mechanical) sciences rather than modern Burkean vitalism.

Topographia

The breadth of Pennant's enterprise in *Arctic Zoology* cannot be underestimated, as the author combines zoology with topography to describe a curious landscape of wild but 'innocent' creatures alongside craggy, volcanic structures projecting from seas and formed into enormous caves. Drawings in the National Library of Wales by Moses Griffith, presumably undertaken for Pennant's *Tour in Wales*, 1778–81, bear a strong similarity to those by the Revd Charles Cordiner (1746–94), which were published in *Antiquities and Scenery of the North of Scotland, in a series of Letters to Thomas Pennant, Esqr.* (1780), and some were also selected by Pennant

Figure 10.4 Moses Griffith, *Glyder-fach, Snowdonia* (undated), graphite and watercolour on paper.

for publication in *Arctic Zoology*. The drawings express a sense of visual dissonance as the artists attempt to depict the precipitous coasts and inland 'Alps' of Wales and Scotland. While these northern voyages, and in particular Joseph Banks's 'discovery' of Staffa, confirmed in the travellers' minds the Vulcanist origins of the world, Griffith's and Cordiner's drawings attempt to render these theories pictorially.[57] One drawing, attributed to Griffith, shows the summit of Glyder-fach, Snowdonia (Fig. 10.4) as a sharp, geometric series of jagged rocks projecting into mist and cloud. The effect is not sublime or picturesque but it does capture something of the artist's uneasiness in this foreboding landscape.

Where this incredible rock-faced 'natural architecture' appears most striking is in Cordiner's depiction of the Doreholm in Shetland and the Bullers of Buchan (Fig. 10.5) off the coast of Aberdeen:

> Bullers of Buchan, and the noble arched rock, so finely represented by the pencil of the Reverend Mr. Cordiner, are justly esteemed the wonders of this country. The former is an amazing harbor, with an entrance through a most august arch of great height and length.[58]

Figure 10.5 Revd Charles Cordiner, *Bullers of Buchan* (1769), graphite and watercolour on paper.

For his account of the Doreholm, Pennant quotes from Captain Thomas Preston's *Two Letters Addressed to Mr. Joseph Ames*, 1744: 'such monstrous precipices and hideous rocks as bring all Brobdingnag before your thoughts'.[59]

The drawing to accompany this description of the Doreholm is attributed to the Revd George Low (1747–95), minister of Birsa, Orkney, and reflects the dangerous rocks and steep incline of the massive arch that certainly recalls the fictional landscape of Swift's giants. Confronted with the strained circumstances of the inhabitants of these most distant islands, Pennant is astonished by their extraordinary feats of agility as a means of sustaining life. This account reveals most clearly the interconnections of zoology and topography and the desire of naturalists to understand this most intimate of relationships between living species. Of the 'Feroe' islanders he writes:

> The dexterity of the fowlers is amazing; they will place their feet against the front of the precipice, and dart themselves some fathoms from it, with a cool eye survey the places where the birds nestle, and again shoot into their haunts.[60]

The swiftness and acrobatic abilities of the fowlers in Pennant's description suggest a similarity between the characteristics of the men and the birds, much as William Hunter's account of *The Nilgai* attributed the animal with certain human qualities such as obsequiousness, so Pennant's fowlers are imbued with bird-like deftness. The purpose of Pennant's reverse anthropomorphizing is not intended to relate a simple attribution of animal mannerisms to human beings. Rather, the account seeks to convey a strong awareness of the synergy between living creatures, particularly under perilous conditions. Just as George Stubbs's anatomical correctness relays a visceral experience of the sublime, Pennant's presentation of the extremities of life lived on the cold, hard rock face, delineated in sharp detail in the accompanying drawing, provides the reader with a vitalizing narrative.

Natura

In their pursuit of nature, Thomas Pennant, William Hunter and George Stubbs understood that the work of artists was essential in the production of knowledge of natural history and of objects related to the natural sciences. Their collaborations reflect the concerns of naturalists to combine a number of interrelated subjects, including anatomy, zoology, geology and topography, among many others. The extent of their interests reached as far as the elite environs of the Royal Academy of Arts and the Royal Society and to the furthest points of antiquarian curiosity at the peripheries of the globe. The potential for elaborate and stimulating artworks that would result from such curiosity was combined with an apprehension that images might not be 'truthful' or accurate enough to convey understanding. These concerns are intimated in a letter from the artist and topographer, Paul Sandby to Thomas Pennant in 1777:

> I make no doubt you will do this delightful country justice in the description of it and tho' Mr Wyndham says in a letter to a friend of his in London that it is impossible to do it justice by the pencil of an artist he is I understand taking a tour thro' Wales, and [...] Carries with him a very excellent Draughtsman and is himself a very good one.[61]

Recourse to empiricism demanded by eighteenth-century naturalists was indeterminate; the images subject to such 'distinguishing' eyes reveal the various convergences of debates on the role of the fine arts in the imitation of nature and on the workings of mimesis. Anatomical experimentation constituted just one aspect of the natural sciences that came to influence artists' representation of the natural world, intensifying and exhorting viewers to an understanding

of natural knowledge. This quality of arresting imagery might also be attributed to the depiction of topographical features of the landscape described in
this essay. However, they are, perhaps, most notable in the ways that naturalists
like Pennant, Hunter and Stubbs deliberately selected specific artistic methods
and mediums to convey an expansive view of the natural world.

Notes

1 John Gascoigne, *Joseph Banks and the English Enlightenment: Useful Knowledge and Polite
   Culture* (Cambridge: Cambridge University Press, 1994), 92. The phrase 'an enlarged
   view' is taken from Thomas Falconer's letter to Sir Joseph Banks, 4 February 1772:
   State Library New South Wales, Papers of Sir Joseph Banks, CY3003, 348–51.
2 Pennant and Stubbs also knew William Hunter's brother, John Hunter (1728–1793),
   and Stubbs shared his interest in comparative anatomy. However, this essay focuses on
   William Hunter in his role as first professor of anatomy at the Royal Academy of Arts,
   acknowledging that he was undoubtedly well versed in knowledge of the fine arts.
3 Bernard Smith, *European Vision and the South Pacific* (New Haven and London: Yale
   University Press, 1985), 3. Smith's study focuses on the work of artists who travelled
   with explorations to the South Pacific. However, his definition of an empirical habit of
   vision, developed by artists as a sophisticated method of seeing and recording, rather
   than 'naïve or unselective naturalism', is an important distinction for the purposes of
   this essay.
4 Falconer to Banks, 4 February 1772 (see 1n).
5 The use of Enlightenment aims to frame the work of Pennant, Hunter and Stubbs is
   perhaps best summarized as a combination of ideas from Scottish Enlightenment writers such as Adam Smith and his description of the stadial development of human society in the 1760s, and Lord Kames, *Sketches of the History of Man* in the 1770s (see Geoff
   Quilley, *Empire to Nation: Art History and the Visualization of Maritime Britain, 1768–1829*
   (New Haven and London: Yale University Press, 2011), 8–72), combining with other
   notable characteristics of the English Enlightenment, 'as a set of barely conscious social
   attitudes which coloured the values and actions of society' (Gascoigne, *Joseph Banks*, 34).
6 Thomas Pennant, *British Zoology*, 4 vols (London, 1776–7), I: xvi.
7 John Locke, *An Essay Concerning Human Understanding* (London, 1690); William Harvey,
   De motu cordis, et sanguinis in animalibus, anatomica exercitation (Leiden, 1639).
8 Barbara Maria Stafford, *Body Criticism: Imaging the Unseen in Enlightenment Art and Medicine*
   (Cambridge, Mass and London: MIT Press, 1994), 12.
9 Sam Smiles, *Eye Witness: Artists and Visual Documentation in Britain, 1770–1830* (Aldershot:
   Ashgate, 2000), 41.
10 Stafford, *Body Criticism*, 44. Nevertheless, such 'somatic metaphors' between landscape
   and the body created tensions between naturalists' desire for images that produced and
   informed knowledge of the natural world and those images that appealed to an earlier
   connoisseurship of the picturesque. For example, 'One of the tasks of the new science
   was held to be the recovery of a "natural" language of description purged of metaphor
   or allusion'; see Richard Hamblyn, 'Private Cabinets and Popular Geology: The British
   Audience for Volcanoes in the Eighteenth Century', in *Transports: Travel, Pleasure, and
   Imaginative Geography, 1660–1830*, ed. Chloe Chard and Helen Langdon (New Haven
   and London: Yale University Press, 1996), 202.

11 See especially Lorraine Daston and Peter Galison, *Objectivity* (New York: Zone Books, 2007). For an understanding of the value of the collation and circulation of knowledge in literary and visual forms, particularly as this relates to geographical imagination, the construction of national identity and empire, see Quilley, *Empire to Nation*; idem, 'Mapping the Art of Travel and Exploration', *Journal of Historical Geography* 43 (2014): 2–8; Smiles, *Eye Witness*; Gascoigne, *Joseph Banks*.

12 Thomas Pennant, *The Literary Life of the Late Thomas Pennant, Esq., by Himself* (London, 1793), 1.

13 Idem, *British Zoology*, I: vi.

14 Ibid. I: x.

15 Ibid. I: xvi.

16 Ibid.

17 The emergence of an 'academic' school of painting in opposition to a 'naturalist or natural' school is a well-established idea in the history of British art of the period, originating in Joshua Reynolds's disapproval of the naturalist quality of Dutch painting, an approach he adopted from reading earlier theoretical treaties, such as Jonathan Richardson (the Elder), *An Essay on the Theory of Painting* (1715); see Sir Joshua Reynolds, *A Journey to Flanders and Holland*, ed. Harry Mount (Cambridge: Cambridge University Press, 1996). This perceived antagonism has persisted in subsequent histories of art in Britain, though recent research has indicated a more nuanced interpretation of teaching at the Royal Academy of Arts in its early years; see Sarah Monks, John Barrell and Mark Hallett, eds., *Living with the Royal Academy: Artistic Ideas and Experiences in England, 1768–1848* (Farnham, Surry and Burlington VT: Ashgate, 2013).

18 Pennant, *British Zoology*, I: xvii.

19 Penny Fielding, *Scotland and the Fictions of Geography, North Britain, 1760–1830* (Cambridge: Cambridge University Press, 2008), 145.

20 *Lloyd's Evening Post*, 1 January 1769.

21 In October 1769 William Hunter delivered his first set of lectures to students at the Royal Academy of Arts; see Royal Academy Archives, London, C/17.

22 Thomas Pennant, *Synopsis of Quadrupeds* (Chester, 1771), x. Pennant was criticized in his own time for his dependence on others' research, most famously, perhaps, by Dr Percy, as recalled by James Boswell and published in the *Critical Review*, January 1772. As Paul Evans shows in this volume, Horace Walpole was also sceptical of Pennant's claims to empiricism. See Charles J. Withers's introduction to *A Tour of Scotland and Voyage to the Hebrides, 1772* (Cambridge: Cambridge University Press, 2000), xiv.

23 William Hunter, 'An Account of the Nyl-ghau, an Indian Animal not hitherto described', *Philosophical Transactions of the Royal Society of London* LXI (1771): 170–81. In these researches, William was also assisted by his brother, John Hunter.

24 Ibid. 170.

25 The word 'antiquarian' perhaps best captures the ways in which the work of all three men is interrelated. Smiles, *Eye Witness*, 11, notes of various antiquarian pursuits: 'One of the things that might bind such disparate activities together might be their participation in the realm of useful knowledge, aided and abetted by the inclusion of visual images.'

26 Hunter, 'An Account', 179. A footnote here adds: 'Mr. Pennant, whose love of natural history heightens the enjoyment of an independent fortune, in his *Synopsis*, published since this paper was written, classes this animal (White-footed, p. 29) as a species of the Antelope; but he now thinks it belongs to another Genus, and will class it accordingly in his next edition.'

27 Sir Joshua Reynolds, *A Discourse to Students at the Royal Academy of Arts*, 14 December 1770.
28 Harry Mount, 'Jan van Rymsdyk and the Nature-Menders: An Early Victim of the Two Cultures Divide', *British Journal for Eighteenth-Century Studies*, 29 (2006): 79–96.
29 John and Andrew van Rymsdyk, *Museum Britannicum* (London, 1778), v.
30 Ibid. iv.
31 Judy Egerton, *George Stubbs, Painter* (New Haven and London: Yale University Press, 2007), 69–71, explains the circumstances which led to Stubbs's disagreement with the Royal Academy. Despite Tobias Smollett's claim that: 'To investigate the life of the animal, is a study becoming a philosopher', the designation 'animal painter' implied a lesser status than 'history painter': *Critical Review, Or the Annals of Literature* 16 (October 1763).
32 See Martin Kemp, *Dr William Hunter at the Royal Academy of Arts* (Glasgow: University of Glasgow Press, 1975). See also note 28 above. Harry Mount, in his introduction to Reynolds, *A Journey to Flanders and Holland*, describes the president's reappraisal of Dutch art, which he had previously referred to as of a 'lower order'.
33 Smiles, *Eye Witness*, 8.
34 Ibid. 40.
35 Thomas Pennant, *A Tour in Scotland, and Voyage to the Hebrides, MDCCLXXII* (Chester, 1774), iii.
36 These included John Lightfoot, author of *Flora Scotica* (1777), and Sir Joseph Banks. Pennant gives a full list of acknowledgments in the advertisement to the first volume.
37 Withers, 'Introduction', xxii. See also Fielding, *Scotland and the Fictions of Geography*; Fredrik Albritton Jonsson, *Enlightenment's Frontier: The Scottish Highlands and the Origins of Environmentalism* (New Haven and London: Yale University Press, 2013).
38 Pennant, *A Tour in Scotland […] 1772*, I: 169–70.
39 Ibid.
40 Banks Correspondence, DTC 1, f. 42, Thomas Falconer to Joseph Banks, 16 January 1773, quoted in Smith, *European Vision*, 14.
41 John Bonehill, '"New scenes drawn by the pencil of Truth": Joseph Banks's Northern Voyage', *Journal of Historical Geography* 43 (2014): 9. See also Nigel Leask, *Curiosity and the Aesthetics of Travel Writing, 1770–1840, 'From an Antique Land'* (Oxford and New York: Oxford University Press, 2002).
42 Pennant, *British Zoology*, I: 124–25.
43 Ibid. 45.
44 Ozias Humphry, 'Particulars of the Life of Mr. Stubbs', transcribed in Nicholas J. Hall, ed., *Fearful Symmetry: George Stubbs, Painter of the English Enlightenment* (New York and London: Hall & Knight, 2000), 201.
45 Stubbs's work has often been misunderstood in this respect. For example, his last unfinished project, *The Comparative Anatomical Exposition of the Structure of the Human Body with that of a Tiger and a Common Fowl* (1795–1806), explores the physiology of humans with less complex animals and appears to follow the type of experimental comparative anatomy carried out by John Hunter in his *Natural History of the Teeth* (London, 1788); see Malcolm Cormack, 'Stubbs and Science', in *Fearful Symmetry*, ed. Hall, 45; Martin Myrone, *George Stubbs* (London: Tate Publishing, 2002).
46 See Aris Sarafianos, 'Stubbs, Walpole and Burke: Convulsive Imitation and "Truth Extorted"', *Tate Papers* 13 (2010): 10–11; idem, 'The Contractility of Burke's Sublime and Heterodoxies in Medicine and Art', *Journal of the History of Ideas* 69 (2008): 23–48; Peter Reil, *Vitalizing Nature in the Enlightenment* (Berkeley and Los Angeles: University of California Press, 2005); Ludger Schwarte, 'The Birth of Aesthetics from the Spirit of Experimentalism', *Proceedings of the European Society for Aesthetics* 1 (2009): 20.

47 Sarafianos, 'Stubbs, Walpole and Burke'. Schwarte, 'Birth of Aesthetics', 21, argues for the significance of anatomical experimentation at the very origins of aesthetics.
48 William Hunter, University of Glasgow Special Collections, Hunter H150, 'Various Notes on the Orignal – Canadian name for the Moose'; Ian W. D. Rolfe, 'William Hunter on the "Irish Elk" and "Stubbs' Moose"', *Archives of Natural History* 11 (1983): 263–90.
49 James Thomson, 'Winter' (1726), *The Seasons by James Thomson to which is prefixed an Essay on the Plan and Character of the Poem by John Aikin* (London, 1778), 221. Pennant quotes the last four lines from this citation.
50 Thomas Pennant, *Arctic Zoology* (London, 1784–85), 17.
51 Ibid. 21.
52 Ibid. iv. Aikin's support for Thomson's *The Seasons* has been attributed to a 'realigning of European historical geography from a Southern to a Northern empire of knowledge', of which Pennant's Scottish tour is undoubtedly a leading constituent. See Stephen Daniels and Paul Elliott, '"Outline maps of knowledge": John Aikin's Geographical Imagination', in *Religious Dissent and the Aikin Barbauld Circle, 1740–1860*, ed. Felicity James and Ian Inkster (Cambridge: Cambridge University Press, 2014), 96.
53 Aikin, *The Seasons by James Thomson*, vi.
54 Ibid. xi.
55 Ibid. xxxvi.
56 Pennant, *Arctic Zoology*, iv.
57 See Hamblyn, 'Private Cabinets'.
58 Pennant, *Arctic Zoology*, xviii.
59 Captain Thomas Preston, 'Two Letters from Mr. Thomas Preston to Mr. Joseph Ames, FRS, Concerning the Island of Zetland', *Philosophical Transactions of the Royal Society of London* VIII (1744): 59.
60 Pennant, *Arctic Zoology*, xli.
61 NLW 14005E, Paul Sandby to Thomas Pennant, 23 June 1777. Henry Penruddocke Wyndham (1736–1819) published his *Tour Through Monmouthshire and Wales* in 1781 and was accompanied on the journey by the Swiss artist Samuel Hieronymus Grimm (1733–1794). Pennant also recommended Sandby's own views of Wales, 'in whose labours fidelity and elegance are united', to his readers in the introduction to *A Tour in Wales*, 1778. See also John Bonehill and Stephen Daniels, eds., *Paul Sandby: Picturing Britain* (London: Royal Academy of Arts, 2009).

Chapter 11

PENNANT'S LEGACY: THE POPULARIZATION OF NATURAL HISTORY IN NINETEENTH-CENTURY WALES THROUGH BOTANICAL TOURING AND OBSERVATION

Caroline R. Kerkham

Introduction

In the summer of 1773 Sir Joseph Banks (1743–1820) and his friend the Revd John Lightfoot (1735–1788) embarked on a botanical expedition through south and north Wales, originally designed to incorporate John Ray's seventeenth-century Welsh itineraries of 1658 and 1662, but in reverse order.[1] They were joined at points along the way by botanists – Holcombe, Skinner, Williams and Davies – Davies being the Revd Hugh Davies, FLS (1739–1821), who later published *Welsh Botanology*, a county flora of Anglesey, the first of its kind for Wales.[2] The Revd John Holcombe (1710–1775) of Pembrokeshire was a friend of Lightfoot who recorded plants in his tour journal which were originally identified by Holcombe.[3]

As a botanical correspondent of Banks and the Revd Sir John Cullum, FRS, FSA (1733–1785), Holcombe may have unwittingly instigated the Pembrokeshire stage of Banks's and Lightfoot's journey through his descriptions of local flora.[4] John Cullum's brother, Sir Thomas Gery Cullum (1741–1831), also a botanist and antiquary, undertook south and mid-Wales tours in 1775 and again in 1811.[5] Thomas Cullum described in his journals the religious sects, local customs, estates, language and state of the poor. In 1775 he climbed Cader Idris and visited 'Pistill Raidr' (Pistyll Rhaeadr), near Llanrhaeadr-ym-Mochnant, and noted between Dinas Mawddwy and Dolgellau 'neat little and well stocked [cottage gardens] [...] Potatoes much cultivated in them.'[6] Stackpole Court's hothouse produce was praised in the later tour, while yew trees in Llanspyddid churchyard were considered to be 'of uncommon magnitude'.[7] Cullum's empirical observations demonstrated a growing trend in the way that information, whether of an antiquarian, natural historical or topographical nature was recorded in travellers' diaries and journals in the eighteenth century.[8] This essay aims to view the way this trend impacted upon botanical exploration in the nineteenth century, leading to a greater emphasis on the geographical distribution of plants, which in turn led to a greater ecological awareness of environmental vulnerability.

Banks's 1773 journey into Wales was not, however, his first Welsh tour: in August 1767 he had travelled into Wales, staying with his relative Robert Banks Hodgkinson at Edwinsford, Carmarthenshire, and concluded his tour by visiting Thomas Pennant at Downing. Banks's journal commented upon local botany, antiquities (including a visit to the caves at Dolaucothi), geology and local agriculture.[9] He also remarked upon 'two Cedars of Lebanon, of nearly equal size, the largest 15 feet in height', planted by Hodgkinson.[10] Thomas Pennant began exploring Snowdonia in the 1750s, not specifically for botanical reasons, but influenced by a broader range of natural history and antiquarian interests. As Paul Evans has noted, it was geology which first drew him to the north Welsh mountains.[11] Despite disclaiming any pretence as a botanist, Pennant recorded a number of Snowdon's plants referring to 'Glyder Vawr' and the 'dire waters of Llyn Idwal' as 'of great note among the botanists for rare plants'.[12] This mountainous region, the 'British Alps' as it was described by Thomas Johnson (1600–1644), the first of the great botanical explorers in Wales, attracted the attention of some of the finest scientific and most intrepid plant hunters of the later seventeenth and early eighteenth centuries: John Ray (1627–1705), Edward Lhuyd (1660–1709), Dr Richard Richardson (1663–1741), J. J. Dillenius (1684–1747) and Samuel Brewer (1670–1743).[13]

Pennant undertook many topographical tours throughout Britain over the next three decades, engaging, at the same time, in meticulously organized local information-gathering exercises.[14] A well-connected gentleman, although not fluently Welsh-speaking, Pennant was ideally situated to bombard clergymen and fellow gentry throughout north Wales with questionnaires on antiquities and natural history and to network with some of the most important naturalists of the eighteenth century, including the Revd Dr W. C. Borlase (1696–1772), Joseph Banks, Carl Linnaeus (1707–1778) and Sir James Edward Smith (1759–1828).[15] Following Edward Lhuyd's practice, including that of requesting information on 'curiosities', examples of Pennant's questionnaires were published in *The Scots Magazine* (1772) and appended to *A Tour in Scotland, MDCCLXIX*.[16] Pennant corresponded with Gilbert White (1720–1793), as did the antiquary and naturalist, Daines Barrington (1727/8–1800), which led eventually to the publication in 1788 of White's *The Natural History and Antiquities of Selborne*. Richard Mabey in his biography of Gilbert White reflected that Pennant, unlike White, was not a natural fieldworker but was 'essentially an intellectual entrepreneur, a popular-izer and compiler of other people's observations and ideas […] but […] always an innovator as far as ideas for books were concerned'.[17] Pennant's reputation as a naturalist was established by the publication of *British Zoology* (1766) and with *The Synopsis of Quadrupeds* (1771), which was republished in 1781 as the *History of Quadrupeds*. Writing in *British Zoology* at a time of growing interest in natural his-tory in Europe, he advocated that 'the *British* reader [should examine] his native riches' as had Linnaeus for Sweden and directed naturalists to Caernarfonshire's 'rich variety of uncommon vegetables'.[18]

Travel and the Systematic Study of the Distribution of Plants

'With the return of Banks and Lightfoot to England (in August 1773)', wrote R. Gwynn Ellis in *Plant hunting in Wales*, 'the last of the great plant-hunting expeditions to Wales came to an end.'[19] He argued that by the beginning of the nineteenth century, the drive to discover new plants throughout Britain was exhausted and what emerged instead was the systematic study of the distribution of plants, culminating in Hewett Cottrell Watson's *Topographical Botany* (1873–74). Influenced by the explorer and naturalist Alexander von Humbolt (1769–1859) and his writings on the interrelationship of physical sciences determining the location of specific plants, by the 1830s Watson had become interested in the concept of geographical botany.[20] *Topographical Botany*, modelled on Watson's four-volume work *Cybele Britannica* (1847–59), divided Britain into vice-counties – a system botanists use to this day – Wales being separated into 13 vice-counties, vc35 and vc41–vc52.[21] The success of Watson's study was dependent upon the skill of innumerable botanists, ama-teur and professional, who traversed the countryside documenting and col-lecting plants which were diligently entered on Watson's plant lists entitled *The London Catalogue of British Plants*, published by the Botanical Society of London. Listed systematically, these catalogues enabled Watson to subject plants and accompanying specimens to rigorous species identification thus furthering geographical distribution studies.[22]

The Post-1773 Botanizing Adventurers: Their Approach to the Natural Landscape

The combination of observations provided by Pennant's *Tour in Wales*, on antiq-uities, agriculture, geology, botany, customs and manners, the landscape and estates, proved to be a heady mixture for travellers, soon to be confronted by Continental travel restrictions caused by the ensuing French and Napoleonic wars. Pennant's tour formula was to be embraced and reinvented by tourists, whose travel accounts were published from the late eighteenth century and over the next hundred years.

Alongside fact-finding missions popularized by Pennant, in the decades following the 1770s there developed in travellers a greater understanding of landscape aesthetics, encapsulated in the writings of the Revd William Gilpin (1724–1804).[23] Ideas expressed in Gilpin's picturesque *Tours* – first published in the 1780s, although originally undertaken in the 1770s – had a major impact upon landscape appreciation, particularly those expressed in *Observations on the River Wye* (1782), and those in his treatise *Three Essays: on Picturesque Beauty; on Picturesque Travel; and on Sketching Landscape* (1792). Consequently, his views on the picturesque influenced landscape writing over the next century. An examination, for example, of the topographical tours of Arthur Aikin (1773–1854), John Evans (1768–*c.*1812), William Bingley (1774–1823) and Benjamin Heath Malkin (1769–1842), and the agricultural and horticultural writings of Arthur Young (1741–1820) and Walter Davies (Gwallter Mechain; 1761–1849), reveals that all were capable, when occasion demanded, of viewing the landscape through picturesque eyes.[24] Pennant himself was not unmindful of landscape aesthetics.[25] Arthur Young's description of Persfield (later Piercefield) on the Wye, compares well with some of the best later picturesque writings. However, his attributing 'the *sublime* to the amazing objects at Persfield' – he described the 'bold hands of the genius of the place' as having created points along the precipice walk which could 'be truly called […] full of the terrible sublime' – and contrasting them with the range of 'beauty' found at Hamilton's Painshill, Surrey, firmly establishes Young's empirical approach as founded in Edmund Burke's essay *A Philosophical Enquiry into the Origin of our Ideas of the Sublime and the Beautiful* (1757).[26] Evans's north-Welsh tour of 1798, with its avowed purpose of 'Botanical Researches in that Alpine Country', considered that the traveller as a 'man of feeling' would be subsumed by curiosity for what he observed; declaring the country until recently 'almost inaccessible' and 'in a great measure, unexplored'. While elevating botanists and antiquarians above 'the superficial traveller' who viewed the landscape purely from the standpoint of 'taste', his enchantment with Powis's parkland was nevertheless described in picturesque terms.[27]

William Bingley, FLS (1774–1823), clergyman, ardent publicist and faithful follower of Pennant's tour format, pertinently remarked that Pennant 'was the first who made a taste for home travels so prevalent in this country'. Somewhat critical of Pennant's indexing, Bingley consequently assiduously annotated and footnoted the two-volume editions of his own *Tours*, including no less than 102 references to Pennant or his *Tour in Wales* throughout the text.[28] However, like his fellow cleric John Evans, Bingley's passion for information did not preclude embracing Gilpin's popular aesthetic. He was captivated by the deep Tryston valley which led him to Rhaeadr Cynwyd, near Corwen, where the waterfall presented wild and romantic confusion in a scene of 'picturesque beauty'.[29] Echoing Pennant's response to Cwm Idwal, 'a fit place to inspire murderous thoughts, environed with horrible precipices, shading a lake', Bingley found the valley 'gloomy and dismal', but nevertheless viewed its 'horrid chasm' of Twll Du with 'sublime pleasure'.[30] Like many of the later tourists, Pennant used the language of landscape appreciation to clothe facts in an agreeable picture, particularly for 'the traveller who delights in wild nature'.[31]

In the 1830s taxonomist C. C. Babington (1808–1895), professor of botany at Cambridge from 1861 until his death, wrote appreciatively of the landscape in his journal in which he recorded visits to north, south and mid-Wales between 1830 and 1883. Two excursions into north Wales in 1830 and 1832, in particular, reveal an interesting sensitivity to atmosphere. On the afternoon of 4 September 1830 he climbed Snowdon and later commented, after collecting plants from the copper mine precipice and ascending the top in clear weather, that soon 'clouds collected around the mountain, but quite under our feet, so that we saw a complete sea of clouds, which was the most beautiful sight that I ever beheld'.[32] Babington, an inveterate traveller throughout Britain, was invariably accompanied on his excursions by fellow naturalists: in 1835, he joined William Borrer, FLS (1781–1862), botanist and lichenologist, on a brief summer tour of north Wales. Babington had first met Borrer in 1834, at Bath, when Borrer gave him 'a number of plants gathered by him in Wales'. Borrer's botanizing tours were sufficiently well known by 1857 for William Pamplin, ALS (1806–1899) to request a mid-Wales itinerary from him.[33] Babington's Welsh interests broadened and he became closely involved in the Cambrian Archaeological Association from 1850 to 1885.[34]

Touring and Botanical Observation: The Role of Societies and Field Clubs

Displaying an increasingly polymathic curiosity and enthusiasm for their surroundings, the late eighteenth-century traveller and nineteenth-century counterpart recorded observations in diaries, letters and journals, many of which were later published. They made drawings in sketchbooks, produced paintings and, from the mid-nineteenth century onwards, took photographs, and from 1894 with the Post Office's blessing, bought postcards.[35] Despite women's growing competence in scientific study and practice by the nineteenth century, nevertheless, access to professional institutions was largely barred to them; the exceptions being the Zoological Society and Botanical Society of London which actively encouraged women members from their inception.[36]

However, for all botanists and other natural historians, the establishment throughout the nineteenth century of societies, magazines, journals and natural history societies and field clubs, provided ideal vehicles for information and ideas exchange, including material on Wales. Price W. Carter, in groundbreaking articles on the history of botanical exploration in Wales recorded many of the local field clubs' excursions which added significantly to plant distribution records in the nineteenth and early twentieth centuries.[37]

Of the national societies, the most influential after the Linnean Society of London (1788) and the Wernerian Natural History Society founded in Edinburgh in 1808, were the Geological Society of London (1807), the Zoological Society (1826), the Botanical Society of London (1836) and its mid-century successor, the Botanical Exchange Club, and the Botanical Society of Edinburgh (1836).[38] J. C. Loudon's two journals, *The Gardener's Magazine* (1826–43) and the *Magazine of Natural History* (1828–36), and *The Phytologist* (1844–63) established by Edward Newman, FLS (1801–76), printer, publisher and botanist, encouraged popular interest in natural history and were important vehicles for the circulation of research and exploration, by amateur as well as by increasingly specialist naturalists. Women naturalists often used journals as a means of publishing their discoveries, albeit occasionally incognito. Newman received an anonymous letter in 1844 from 'A lady who has this season visited North Wales' where she found 'Asplenium septentrionale in great abundance, and also unmistakable specimens of Asplenium germanicum'; the forked spleenwort and alternate leaved spleenwort. Uncertain that his correspondent was familiar with the latter plant, Newman waited for confirmation of a specimen from the same location 'by Mr W. Wilson, of Warrington'.[39] William Wilson (1799–1871) was a well-known bryologist: William and Caroline Pamplin encountered Wilson at Llanberis when both parties were on north-Welsh walking tours in July 1854.[40] *The Phytologist* continued from 1855 until 1863 under the umbrella of Pamplin's Soho botanical publishing business.[41]

By the 1860s the significant organizations for disseminating information relating to Wales, particularly through publications and fieldwork, appear to have been the Cambrian Archaeological Association (1846), the Powysland Club (1867) and the Cardiff Naturalists Society (1868), with cross-border input from Herefordshire's Woolhope Club (1851) and the Worcestershire Naturalists Club (1847).

Regular field meetings became the norm for most societies by 1900 with a summer Annual Field Meeting as a highlight. The resulting reports from these excursions offer an interesting insight into the extent to which natural history, and botany in particular, had become a serious occupation, largely for the middle and professional classes. The Woolhope Club, for example, from the late 1860s held frequent 'Field-Day' meetings across the border; Llandrindod and Aberedw's environs were explored in 1867; 'the Radnorshire Hills' in 1879, while Rhayader and Cwm Elan were visited in 1881, 1896 and 1899. Exploring continued with field trips to Brecon and the Beacons in July 1882 and to Radnorshire again in 1888 and 1889. Members also used their particular society 'Proceedings' as vehicles for publishing private investigations into 'the practical study of natural history' in their county and adjacent districts, which included collecting as well as recording plants.[42] The Revd Augustin Ley (1842–1911), president of the Woolhope Naturalists Field Club in 1881 and vice-president in 1889, published several articles in the society's *Transactions* on his botanical explorations in Breconshire and Radnorshire between 1879 and 1898.[43]

Botanizing Travellers

From the early nineteenth century tourists of all persuasion – whether scientific, artistic or literary – increasingly came from a broad spectrum of the growing professional middle classes, as well as from the gentry and aristocratic circles. Plants were first studied for themselves rather than for their medicinal values in the seventeenth century and while botany as a science had become established in the universities by the late eighteenth century, the Apothecaries Act of 1815 demanded that anyone studying medicine was required to obtain the LSA (Licentiate of the Society of Apothecaries) qualification, which upheld exacting standards of plant recognition. Sir James Edward Smith, founder and president of the Linnean Society of London, a scion of the Dissenting merchant class in Norwich, used his University of Edinburgh medical degree to train as a botanist.[44] Writing to fellow botanist Thomas J. Woodward (1745–1820) on first visiting Hafod, Cardiganshire, in August 1795, Smith expressed his disappointment that while the landscape presented rugged 'rocks woods and cascades' on a grand scale, the plants observed in Wales were common ones 'of hilly, not alpine countries'. Four days later he added that 'There is no limestone chalk or gravel here, and consequently the Flora is poor.' He mentioned an abundance of golden-rod, saw-wort and imperforate St John's-wort, but little else that was uncommon as well as 'few ferns; many mosses and algae'.[45] However, years later in his *Tour to Hafod*, he expanded on the variety of cryptogams found in Hafod's woods. Using *English Botany* as his reference point, Smith recorded that: 'Buxbaumia foliosa [*Diphyscium foliosum*], English Botany, t. 329, grows on the rocks. The trees are mantled with the most magnificent leafy Lichens, as glomuliferus [*Lobaria amplissima*], t. 293, laete-virens [*Lobaria laetevirens*], t. 294, and scutatus [*Lichen scutatus*], t. 1834.' He enthused:

> Under the wood, on the left bank of the river [Ystwyth], grows a vast profusion of Lichen sylvaticus [*Sticta sylvatica*], among which I have spent many an idle hour in vain hunting for the shields. A narrow path to the right, after crossing the bridge, leads to a close thicket of very ancient thorns, where L. Fuliginosus [*Sticta fuliginosa*], Engl. Bot. T. 1103, and limbatus [*Sticta limbata*], t. 1104, are found in great perfection, but the former only in fructification.[46]

The 36 volumes of *English Botany* by James Sowerby and James Edward Smith were published between 1790 and 1814: Smith produced the scientific descriptions for the 2,592 plates illustrated by Sowerby.[47] Interestingly, out of the 26 plants cited in Turner and Dillwyn's *Botanist's Guide* for Cardiganshire, 20 are associated with Smith's Hafod connection. As late as 1894 J. H. Burkill and J. C. Willis were to comment on the paucity of botanical literature for the county.[48] This despite the attraction of tourists to mid-Wales's picturesque delights at Hafod and the sublimity of the Devil's Bridge, particularly between 1780 and the 1840s.[49]

Facility and opportunity were essential components for botanical touring and modes of travel from the seventeenth century to the late nineteenth century varied from walking – the ubiquitous pedestrian tour – to riding by horse, carriage, or by stagecoach and latterly by railway and bicycle. William Bingley attached an itinerary to his revised *Tour*, entitled a 'Guide to North Wales', detailing where post-chaises or horses could be hired at inns en route.[50] 'His mode of travelling was chiefly on foot', he explained, which was advantageous to the naturalist who could 'strike out of the road, amongst the mountains and morasses […] independent of all those obstacles that inevitably attend the bringing of carriages or horses' and emphasized that the 'traveller of taste […] the naturalist, and the antiquary, have all, in this romantic country, full scope for their respective pursuits'.[51]

In her survey of tourists in Britain, Esther Moir considered that as access and opportunity for cheap excursions reached the masses, the sense of travel 'as a venture of personal discovery' lost much of its distinctiveness.[52] However, a close reading of the literature would suggest that it was still possible for those journeying by train, certainly until the 1880s, to consider themselves 'pedestrian tourists': once reaching their destination they simply walked the landscape. William Pamplin, the London botanical agent and publisher/bookseller, walked north Wales from the 1840s, travelling via Euston or Paddington to Oswestry, or later, from the 1850s, to Llandudno Junction. Pamplin, accompanied by his wife, took advantage of the London and North-Western Railway cheap 'Excursion' trains run by the entrepreneurial agent, Henry R. Marcus.[53] Pamplin climbed Snowdon seven times and was well 'acquainted with the habitats of the rare plants of Carnarvonshire'. He climbed Snowdon accompanied by the Llanberis guide William Williams in July 1857 when they collected a number of ferns, but were careful to take only two fronds of the woodsia.[54] Williams died tragically in 1861, when he fell descending Snowdon.[55]

Edwin Lees, FLS (1800–1887), for his excursion to the Breidden in the late 1840s, hired a gig from Welshpool, leaving it at the Plough and Harrow while he and his guide clambered over the hill. His account of this botanical ramble is a fine essay in landscape writing: Lees conveys, not only his passion for the plants at his feet, but his great joy in the natural world. Halfway up the Breidden, having waxed eloquently about the spiked speedwell, wild marjoram and the common rock-rose which met his eye, he threw himself 'upon the turf, and resting from my up-hill work, for some time contemplated the scene with rapt enjoyment'.[56] A founder of the Worcester Natural History Society in 1833 and in 1847 one of three founders of the Worcester Naturalists Club, Lees – a printer and bookseller by trade – was a prolific writer, commentator and popularizer of natural history.[57] He gave a retrospective account, not dissimilar to Bingley's own experience, of his unnerving climb in Twll Du, Cwm Idwal, in the 1830s or 1840s looking for '*Polystichum Lonchitis* [holly fern], as well as the almost eradicated *Woodsia*', where he notes his guide's cure for vertigo: '"Look up," said he, "creep close to the rock, and there is no danger."'[58]

The Impact of Pteridomania

Reference to the woodsia (Fig. 11.1) brings us to 'pteridomania', or the 'Victorian fern craze'.[59] Techniques for fern cultivation had improved during the early years of the nineteenth century, but by the 1840s and 1850s, recording, collecting and writing about ferns in the wild was all the rage. This, understandably, had dire consequences for ferns in their natural environment. Lees, in a footnote to his plant hunting in Twll Du remarked:

> I am glad to understand that *Woodsia ilvensis* [oblong woodsia] has lately reappeared in unusual abundance at its old station, a rock above Llyn-y-Cwn. About a hundred plants were said to be visible in 1849, but [...] not to be gathered without the aid of a ladder.[60]

Writing in *The Phytologist* of a fern-hunting expedition in September 1849, travelling via the Devil's Bridge into Merioneth and Caernarfonshire, William Bennett

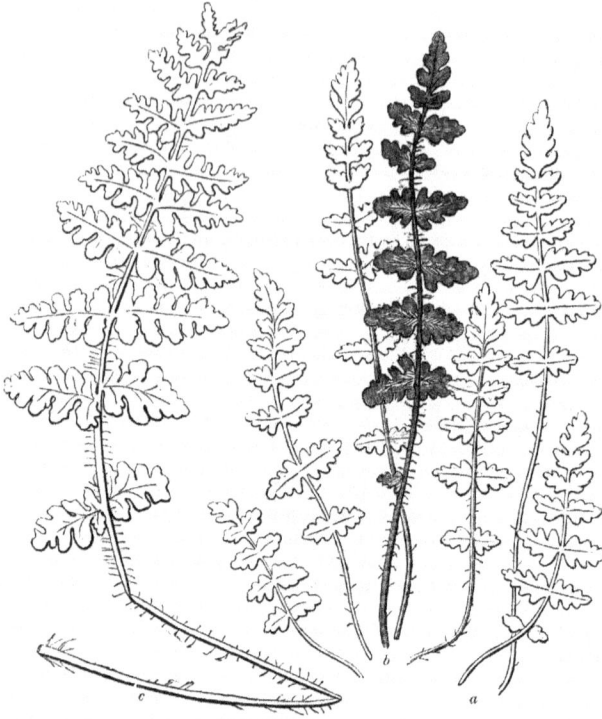

RAY'S WOODSIA (*natural size*).
WOODSIA ILVENSIS, Smith,* Hooker, Babington.

THIS is one of the rarest of our ferns : it roots in the fissures
of rocks, in the most bleak and exposed mountainous regions :
it has hitherto occurred to botanists in two counties in England,
one in Wales, and one in Scotland.

Figure 11.1 *Woodsia Ilvensis,* from Edward Newman, *History of British Ferns,* 2nd ed.
(London, 1844); 'a' illustrates two fronds from Llyn y Cwn given to Newman by Wilson
and Pamplin.

(1804–1873), a Quaker businessman accompanied by his two sons, described fern
habitats in some detail, and also the effort to capture specimens. They had difficulty in collecting a perfect specimen of *Asplenium lanceolatum* (*A. obovatum*; lan-
ceolate spleenwort), one of the spleenworts, as it rooted deeply among rocky

fissures along the road from Barmouth to Harlech. With Newman's *British Ferns* (1840; 2nd ed. 1844) in hand, they searched in vain for the *Asplenium septentrionale* (forked spleenwort) near Llanrwst, concluding that 'some "piratical botanist" had destroyed all plants': however, on nearing Capel Curig, they 'pounced' upon a single specimen growing on a wall. They never did find the woodsia in situ, but their guide posted on 'a small root, and some fronds of a true Woodsia, but [...] so mutilated that we cannot [...] determine the species'.[61] Newman was appalled by the devastation wreaked upon fern habitats through accurate site identification; he blamed himself for the demise of *Asplenium septentrionale* in one station post-1840 and remarked on the pressures faced by William Williams, the Snowdon guide, to reveal the woodsia's location. Commenting in 1854, Newman wrote that the 'utter extermination of these ferns from all accessible places is not only certain, but also imminent'. Unsettled by the problem of over-collecting, he added 'I deeply regret the prevalence of this exterminating spirit, for it tends to deprive the true botanist of one of his greatest pleasures, – that of visiting rare plants in their native localities.'[62] D. E. Allen, in his history of pteridomania and an article on attitudes towards botanical conservation, examines the consequences of Victorian collecting enthusiasm and the drive towards protective legislation.[63]

William Pamplin, as a botanical agent, spent time collecting plants to distribute to clients while on walking tours in north Wales. Borrer, for example, requested Pamplin to find him a specimen of lloydia in 1855 and again in 1857, with no expense to be spared for conveyance.[64] By contrast, in August 1854, Pamplin, having walked from Llanberis nearly to Gorphwysfa and collected a few ferns, encountered the then rector of Llanberis, William Lloyd Williams (1808–73) and recorded an early attempt at conservation:

> we went together into the Ch. Yard and planted such as I had about the Church walls – we find that Asplenium trichomanes [maidenhair spleenwort] has already pretty well established itself upon several parts of the church – note – the S. & E. Wall of the Church Yard might very soon be completely covered with ferns or other plants as it is admirably adapted for introducing them successfully – also on the walls of the Church itself.[65]

The Parson–Naturalists

The tradition of the parson–naturalist reached back to John Ray. As a professional group, clerics, many of whom were Anglican, were notable for their botanical interests. Pennant owed 'the account of our Snowdonian plants' in the *Journey to Snowdon*, to the Revd Hugh Davies, aforementioned author of *Welsh Botanology*.[66] William Bingley attached the *Flora Cambrica*, a 'Systematic Catalogue of the more uncommon Welsh plants', to the second volume of his reworked *Tours*, taken in north Wales in 1798 and 1801.[67] Bingley was probably

Pennant's most faithful disciple; he published works on botany and zoology, but also on musicology and his topographical writings reveal an intensity of research seemingly modelled on Pennant. Pennant's Welsh *Tour* he declared, was not only 'accurate and learned', but also '*out of print* for some time'; perhaps a mild justification for publishing his own *Tour*.[68]

A clamber up Twll Du from Llyn Idwal rewarded Bingley with 'a great number of uncommon plants' among which were alpine meadow-rue, awl-wort, moss campion, an alpine hawkweed and some ferns, mosses and lichens.[69] Bingley's list of plants growing on Snowdon's Clogwyn y Garnedd, in his revised *North Wales* (1804) had, Jones considers, in all probability been given to him by his naturalist companion and guide, the Revd P. W. Williams (1763–1836), rector of Llanrug and Llanberis.[70]

Later in the century, two of the more outstanding taxonomic clerics were the Revd Augustin Ley, vicar of Sellack, near Ross on Wye and the Revd H. J. Riddelsdell (1866–1914). Ley, a botanist, bryologist and, as previously mentioned, a distinguished member of the Woolhope Club, like many other naturalists was an indefatigable walker, recorded in Breconshire between 1874 and 1907. Writing on the Honddu and Grwyne valleys, he declared his love of mountain scenery began at the age of six when he beheld the 'mossy recesses with their miniature waterfalls and pools overtopped by berry-laden Mountain Ashes' in Cwm Bwchel, near Llanthony.[71] He travelled widely throughout Wales from the 1860s, exploring mid-Wales, the Cadair Idris mountain range and parts of Snowdonia.[72] Enthralled by Cwm Idwal and the 'chasm of Twll-du, which affords a view […] unequalled in Wales for wild beauty', he added somewhat optimistically that the cliff was 'in parts, utterly inaccessible; and its (botanical) rarities will never be exterminated by collectors'.[73] Despite growing concern among the botanical community about over-collecting for herbaria, Ley, apart from collecting for his own herbarium, contributed 15,000 specimens to the Botanical Exchange Club during almost a lifetime association with that club. Furthermore, in lamenting his lack of understanding of geology, which he believed would have informed his 'botanical rambles' and study of mosses and liverworts, Ley illustrates the degree to which some late nineteenth-century naturalists were becoming specialists.[74] Botanizing in the Black Mountains and Brecon Beacons, Riddelsdell studied the area around Builth Wells: he later published the *Flora of Glamorgan* in 1907 as a *Journal of Botany* supplement.[75] Two years previously, he published 'Lightfoot's Visit to Wales in 1773', transcribed from Daniel Solander's copy in the Natural History Museum.[76] The polymathic approach to exploration adopted by earlier travellers was gradually being replaced by more terse, systematic reporting – an approach championed by Watson.[77]

Amateurs and Professionals: The Rise of an Ecological Approach to Botanical Recording

Riddelsdell's transcription of Lightfoot's tour was not the first foray into historical botany in Wales: in volume one of the new series of *The Phytologist*, its editor, Alexander Irvine (1794–1873), produced an article on botanical tours in Wales.[78] This covered touring from Giraldus to Edward Lhuyd and Richard Richardson and included a précis (in English) of Thomas Johnson's 'simpling' tour in north Wales of 1639, originally published in Latin in 1641. This was later translated and published without the plant list by W. Jenkyn Thomas.[79]

The precursor to Watson's systematic *Topographical Botany* was Dawson Turner and Lewis Weston Dillwyn's *The Botanists' Guide to England and Wales* published in 1805. Turner (FRS, FLS; 1775–1858) was a Norfolk banker. Dillwyn (FRS, FLS; 1778–1855) was a Swansea businessman from a Pennsylvanian Quaker background, whose Penllergaer estate became a hub of scientific activity in the growing industrial town. In order to produce the *Guide*, they took a leaf out of Pennant's book and circulated the country with a botanical questionnaire.[80] Their publication not only revealed the habitats of plants, but also recorded the names of those travellers who had been and were currently botanizing in the increasingly changing landscape. By the late eighteenth century, agricultural improvement as well as quarrying and mining activity throughout Wales, combined with economic development such as canals and railways, had already impacted considerably on the natural world.[81] It could be argued that the growing focus on plant distribution which brought about a new emphasis on habitat and the plant environment was partly a response to such changes, particularly that of industrial development and creeping urbanization.

In turning their attention to Welsh county plant records, Turner and Dillwyn made it clear that their aim was to encourage other 'Botanists who have leisure and opportunity, to turn their minds to these parts of the kingdom' where little attention had been paid to 'discovery of [nature's] treasures'.[82] The few plants listed for Breconshire, Cardiganshire and the three lichens recorded for Radnorshire spoke for themselves. However, they civilly warned – foreshadowing Watson's heavy-handed approach – that suspect plant identification would mean absence from their listings. Thus John Evans, whose tour in 1798 was a 'Journey, undertaken for Botanical purposes' and who hoped at some future point to publish a '*Flora Cambrica*', found many of his records rejected because local botanists and Turner and Dillwyn themselves were convinced 'that the plants are not always to be found in the stations he has assigned to them'.[83]

In 1841 Edwin Lees published a note on plants from the Aberystwyth district in which he classed plants according to habitat. 'This ecological approach, at such an early date', commented Carter, 'deserve[d] [...] more recognition than it received'.[84] In recording plants on the limestone rocks of the Great Orme in 1849, particularly the elusive *Cotoneaster insculptus*, Lees continued to express an interest in plants' relationship to their environment. Again, his speculation that comparison of contemporary records with those of previous centuries would reveal plant location changes was thoroughly ecological. Although not the only botanical tourist to remark on Llandudno's modernization, Lees's observations give a compelling picture of the infancy of a Victorian seaside town and attendant damage to its native flora. He not only lamented the loss of Llandudno Common's marshland, but the old village and way of life and, only two years later, commented similarly of development at Torquay, Devon. 'Local plants', he observed, 'will soon [...] only exist in herbaria'.[85]

In the summer of 1848, Babington botanized in south Wales: returning to Pembrokeshire in 1851, he recorded in the southern half of the county for several weeks. In his description of the marshlands between Tenby and St Florence and that of Castle Martin, he noted that 'the former [...] never having been well drained, has [...] continued in the condition of rough pasture' and was thus a fair example of marshland vegetation. However, as 'Castle Martin Corse' had been 'completely drained about sixty years since, and much of it converted into arable land', having returned to 'coarse, wet pasture, it is nearly deprived of all its peculiar plants, a few only remaining in the ditches'.[86] In 1857, worried about over-collection, Babington wrote to Pamplin exhorting him not to publish the exact stations of rare plants or ferns, 'for extirpation is the rule in Wales with tourists and collectors who call themselves botanists'.[87] *The Phytologist* duly supported this view, declaring: 'The spread of knowledge is very desirable, but the destruction of scarce plants is a subject to be deplored by every genuine Botanist.' Babington returned to this problem in a letter to the *Journal of Botany* in 1864.[88]

In August 1883 Alexander Goodman More, FLS (1830–1895) toured south-west Wales. More had coauthored with David Moore (1807–1879) the Watsonian-inspired *Cybele Hibernica* (1866). He chronicled unsuccessful, repeated searching for galingale at its locality in Whitesand Bay, Pembrokeshire. Reflecting on the evidence from Banks's and Lightfoot's tour, he commented that 'this very rare species had been gathered just one hundred and ten years ago; [for] an undoubted specimen is preserved in the Banksian Herbarium at the British Museum'. More's concluding observation echoes the prescience of Lees's earlier prediction when he wrote 'I can only conjecture that modern alterations, however slight, have led to the extirpation of the plant.'[89]

Conclusion

The 'Vale of the Elan [...] widely celebrated for its great beauty',[90] despite straddling four counties, lies largely in Radnorshire and was the destination for the Woolhope Club's summer meeting in July 1881. Quoting from W. L. Bowles's poem 'Coombe-Ellen' (1798) and describing the valley as 'the gem of South Wales', the Society's account of the visit evoked 'Scotch or Swiss scenes of singular beauty'. Nevertheless, 'the Woolhope Club [had] ever a scientific object in view, and the botanists were at work in collecting and cataloguing the Radnorshire plants at this time in flower'. Later that day, attenders listened to a lecture on the geology and geography of Cwm Elan given under the shady trees of Cwm Elan House's grounds, 'a highly cultivated oasis in the wild scenery'. The Herefordshire naturalists visited the valley again in 1896 and 1899, by which time those who had been 'familiar with the Elan Valley in its original natural beauty' were confronted by a landscape in radical transition.[91] It was ironic that the building of the Elan Valley Dams for Birmingham's water supply came about because of its geology, purity and softness of water, vast watershed and 'wild moorland' and a local population not exceeding 180 individuals.[92]

In March 1902 Mr G. A. Birkenhead, a committee member of the Cardiff Naturalists Society and civil engineer by profession, read a paper to the Society on his recent survey of the Elan Valley and dam construction works in which he interwove observations on the Elan's landscape and natural history with an account of the river valley, the dam workers' village 'said to be a model' and the dams. While reflecting upon the future Caban Coch reservoir he remarked on the unspoilt nature of the valley: 'we had to literally hew our way through the woods to run our contour and traverse lines [...] Here we are in what may be truly termed Wild Wales.'[93] Birkenhead's observations, while combining natural history and landscape appreciation, also acknowledged the irretrievably changing physical environment.

By the end of the nineteenth century, the dedicated fieldwork of plant recorders in the Welsh countryside had promoted a greater understanding of the plant environment, which in turn contributed to a growing consciousness of the natural world's vulnerability. A handful of prototype *Floras* were collated, some of which were published and formed the basis upon which twentieth-century botanists built their research. More cosmopolitan perhaps than the 'parochial *Geniuses*' Pennant had in mind, these nineteenth-century travellers undoubtedly 'favour[ed] the world with a fuller and more satisfactory Account of their Country'.[94] That botanists, like many other natural historians of the period, were inspired to 'look out' at the wider landscape through which they travelled, is surely a legacy of the information gathering which found voice in Pennant's *Tour in Wales*.[95]

Notes

1 Attracted by Holcombe's description of Pembrokeshire flora, they spent longer than planned in south Wales. Pressure of time forced them to exclude Cadair Idris from their itinerary, reaching Snowdon via the Black Mountains and better roads from Hereford to north Wales. See Harry J. Riddelsdell, 'Lightfoot's Visit to Wales in 1773', *Journal of Botany* 43 (1905): 290–307 (295).
2 Hugh Davies, *Welsh Botanology* (London, 1813).
3 Riddelsdell, 'Lightfoot's Visit to Wales', 296. See Price W. Carter, 'Some Account of the History of Botanical Exploration in Pembrokeshire', *Nature in Wales: The Quarterly Journal of the West Wales Field Society* new series, 5, part 1 (1986): 33–44 (37); Charles C. Babington, 'Pembrokeshire Plants and the Rev. Mr. Holcombe', *Journal of Botany* 24 (1886): 22–23.
4 Riddelsdell, 'Lightfoot's Visit to Wales', 296; Carter, 'Pembrokeshire', 37.
5 NLW 5446B, Thomas G. Cullum, 'Diary of a Journey into South Wales' (1775), 'Journal of a Tour in South Wales' (1811). See Herbert M. Vaughan, 'Tours in Wales, 1775 and 1811', *Y Cymmrodor* 38 (1927): 45–59.
6 NLW 5446B, Cullum (1775): 151–56. He also criticized the condition of Welsh cottages, particularly in Pembrokeshire and Carmarthenshire (125, 150). Cullum was shocked by villagers' poverty he observed in Abergwili, NLW 5446B, Cullum (1811): 91.
7 NLW 5446B, Cullum (1775): 150; (1811): 75, 102.
8 Esther Moir, *The Discovery of Britain: The English Tourists, 1540 to 1840* (London: Routledge & Kegan Paul, 1964); David Thomas, *Agriculture in Wales during the Napoleonic Wars: A Study in the Geographical Interpretation of Historical Sources* (Cardiff: University of Wales Press, 1964).
9 NLW 147C, J. Banks (1767). See George E. Evans, 'Journal of a Tour in Carmarthenshire A.D. 1767', *Transactions of the Carmarthenshire Antiquarian Society* 15, part 37 (1921–2): 14–18, 23–24.
10 Evans, 'Journal of a Tour': 23.
11 R. Paul Evans, 'Thomas Pennant (1726–1798): "The Father of Cambrian Tourists"', *Welsh History Review* 13, no. 4 (1987): 395–417.
12 Thomas Pennant, *A Tour in Wales: The Journey to Snowdon* (London, 1781), 163–64; see Dewi Jones, *The Botanists and Guides of Snowdonia* (Llanrwst: Gwasg Carreg Gwalch, 1996), 64. Pennant's collaboration with Lightfoot over the *Flora Scotica*, published in 1777, was stressful, at least for Lightfoot, because of Pennant's insistence on species description in English, not scientifically accepted Latin. See David E. Allen, *Books and Naturalists: A Survey of British Natural History* (London: Harper Collins, 2010), 108–11.
13 Thomas Johnson, *The Itinerary of a Botanist through North Wales in the Year 1639 A.D.*, trans. W. Jenkyn Thomas (Bangor: Evan Thomas, 1908), 7.
14 Evans, 'Thomas Pennant': 397.
15 See Adam Fox, 'Printed Questionnaires, Research Networks, and the Discovery of the British Isles, 1650–1800', *Historical Journal* 53, no. 3 (2010): 593–621.
16 Thomas Pennant, *A Tour in Scotland. MDCCLXIX* (Chester, 1771); *The Scots Magazine*, 1 January 1772; ibid. 1 April 1772: www.britishnewspaperarchive.co.uk (accessed 21 September 2013). See Allen, *Books and Naturalists*, 104. Anne Secord draws attention to Gilbert White's drawing upon this custom of parochial questions collated by antiquaries for investigating parishes; see Gilbert White, *The Natural History of Selborne*, ed. Anne Secord (Oxford: Oxford University Press, 2013), xv–xvi.

17 Richard Mabey, *Gilbert White: A Biography of the Author of The Natural History of Selborne* (London: Century Hutchinson, 1986), 106.

18 Thomas Pennant, *British Zoology*, 2 vols (London, 1768), I: i–xv (i–ii).

19 R. Gwynn Ellis, *Plant Hunting in Wales* (Cardiff: National Museum of Wales, n.d.), 29; reprints from *Amgueddfa: Bulletin of the NMW* 10 (1972), 13 (1973), 16 (1974).

20 Humbolt's *Essay on the Geography of Plants* was published in German and French in 1807; see Andrea Wulf, *The Invention of Nature: The Adventures of Alexander Von Humbolt, the Lost Hero of Science* (London: John Murray, 2015), 126–30, 431; Nigel Leask, *Curiosity and the Aesthetics of Travel Writing, 1770–1840: 'From an Antique Land'* (Oxford: Oxford University Press, 2002), 246–56; Frank N. Egerton, *Hewett Cottrell Watson: Victorian Plant Ecologist and Evolutionist* (Hants: Ashgate, 2003), 24–37, 83–98.

21 Hewett C. Watson, *Topographical Botany: being the local and personal records towards shewing the Distribution of British Plants traced through the 112 Counties and Vice-counties of England, Wales, and Scotland*, 2nd ed. (London: Bernard Quaritch, 1883), xlvii.

22 David E. Allen, *The Naturalist in Britain: A Social History* (Middlesex: Pelican Books, 1978), 112; idem, *The Botanists: A History of the Botanical Society of the British Isles through a Hundred and Fifty Years* (Winchester: St Paul's Bibliographies, 1986), 34–35.

23 See Carl P. Barbier, *William Gilpin: His Drawings, Teaching and Theory of the Picturesque* (Oxford: Clarendon Press, 1963); Malcolm Andrews, *The Search for the Picturesque: Landscape Aesthetics and Tourism in Britain, 1760–1800* (Aldershot: Scolar Press, 1989).

24 Arthur Aikin, *Journal of a tour through North Wales and part of Shropshire: with observations in Mineralogy, and other branches of Natural History* (London, 1797); John Evans, *A tour through part of North Wales, in the year 1798, and at other times: principally undertaken with a view to botanical researches in the Alpine Country, interspersed with observations on its scenery, agriculture, manufactures, customs, history, and antiquities* (London: J. White, 1800); William Bingley, *A tour round North Wales, performed during the summer of 1798, etc*, 2 vols. (London: E. Williams; Cambridge: J. Deighton, 1800); idem, *North Wales: including its scenery, antiquities, customs, and some sketches of its natural history; delineated from two excursions through all the interesting parts of that country, during the summers of 1798 and 1801*, 2 vols. (London: Longman and Rees, 1804); Benjamin H. Malkin, *The Scenery, Antiquities, & Biography of South Wales, etc* (London: Longman and Rees, 1804); Arthur Young, *A Six Week's Tour, Through the Southern Counties of England and Wales, etc*, 3rd ed. (London, 1772); Walter Davies, *General view of the agriculture and domestic economy of North Wales, etc* (London: B. McMillan, 1813); idem, *General view of the agriculture and domestic economy of South Wales, etc*, 2 vols. (London: McMillan, 1814).

25 Pennant, *Tour in Wales [...] Snowdon*, 161, 171–73.

26 Young, *A Six Week's Tour*, 169–85 (171, 219).

27 Evans, *Tour through part of North Wales*, vi–vii, 11–12.

28 Bingley, *Tour round North Wales [...] 1798*, I: x–xvi; idem, *North Wales [...] 1798 and 1801*, I: v–xii.

29 Ibid. I: 186.

30 Pennant, *Tour in Wales [...] Snowdon*, 162; Bingley, *Tour round North Wales [...] 1798*, I: 210–12.

31 Pennant, *Tour in Wales [...] Snowdon*, 161, 180–82, 184.

32 Charles C. Babington, *Memorials Journal and Botanical Correspondence of Charles Cardale Babington* (Cambridge: Macmillan and Bowes, 1897), 7, 8–15.

33 Ibid. 42, 22; NLW 7507C, William Borrer (1857), Correspondence of William Pamplin, William Borrer to William Pamplin, 22 April 1857.

34 Babington, *Memorials*, lxxxix–xc, 453–54; Nancy Edwards and John Gould, 'From Antiquarians to Archaeologists in Nineteenth-Century Wales: The Question of Prehistory', in *Writing a Small Nation's Past: Wales in Comparative Perspective, 1850–1950*, ed. Neil Evans and Huw Pryce (Surrey: Ashgate, 2013), 143–64.

35 See Maurice W. Barley, *A Guide to British Topographical Collections* (Nottingham: Council for British Archaeology, 1974); Robin Gard, ed., *The Observant Traveller: Diaries of Travel in England, Wales and Scotland in the County Record Offices of England and Wales* (London: HMSO, 1989).

36 See Allen, *Botanists*, 12; Adrian Desmond, 'Redefining the X Axis: "Professionals", "Amateurs" and the Making of Mid-Victorian Biology: A Progress Report', *Journal of the History of Biology* 34 (2001): 3–50 (16–21); Ann B. Shteir, *Cultivating Women, Cultivating Science: Flora's Daughters and Botany in England 1760 to 1860* (Baltimore and London: John Hopkins University Press, 1996), 171–94; Linda Lear, *Beatrix Potter: A Life in Nature* (London: Allen Lane, 2007), 123–24.

37 See, e.g., Price W. Carter, 'Some Account of the Botanical Exploration of Denbighshire', *Transactions of the Denbighshire Historical Society* 9 (1960): 114–45 (142–43). See also Diarmid A. Finnegan, *Natural History Societies and Civic Culture in Victorian Scotland* (London: Pickering & Chatto, 2009), 45–65, for an examination of fieldwork as a recreational pursuit; idem, 'Natural History Societies in Late Victorian Scotland and the Pursuit of Local Civic Science', *British Journal for the History of Science* 38, no. 1 (2005): 53–72.

38 Allen, *Naturalist in Britain*, passim.

39 Edward Newman, 'Welch Habitat for *Asplenium germanicum*', *Phytologist* 2 (1844–7): 974–75.

40 Mark Lawley, 'Bygone Bryologists: William Wilson (1799–1871)', *Field Bryology* 96 (2008): 39–43, http://rbg-web2.rbge.org.uk/bbs/Activities/field%20bryology (accessed 9 May 2016); NLW 7509C, William Pamplin (1854), 2/vi: 'A tour in North Wales in the summer of 1854', 27 July 1854.

41 Allen, *Botanists*, passim.

42 See 'Rules of the Woolhope Naturalists Field Club', *Transactions of the Woolhope Naturalists Field Club* I (1852): v–vi.

43 Price W. Carter, 'A History of Botanical Exploration in Brecknock', *Brycheiniog* 3 (1957): 157–80 (163–65, 168, 178–79).

44 Allen, *Botanists*, 6; Desmond, 'Redefining the X Axis', 35–38; Margot Walker, *Sir James Edward Smith MD, FRS, PLS, 1759–1828: First President of the Linnean Society of London* (London: Linnean Society, 1988), 1–3.

45 Linnean Society, London, Smith Papers: General Correspondence, 18.206–8: J. E. Smith to T. J. Woodward, 28–31 August 1795.

46 James E. Smith, *A Tour to Hafod in Cardiganshire, the seat of Thomas Johnes, Esq. M.P.* (London: White, 1810), 12.

47 Allen, *Books and Naturalists*, 134.

48 Isaac H. Burkill and John C. Willis, 'Botanical Notes from North Cardiganshire', *Journal of Botany* 32 (1894): 4–10; Arthur O. Chater, *Flora of Cardiganshire* (Aberystwyth: privately published, 2010), 15.

49 See, e.g., NLW Microfilm 215, George Cumberland, 'Journal of a Tour in Wales, 1784'; George Cumberland, *An Attempt to Describe Hafod* (London: Egerton, 1796); Smith, *Tour to Hafod*; Thomas Roscoe, *Wanderings and Excursions in South Wales; including the Scenery of the River Wye* (London: C. Tilt and Simpkin, 1837), 15–22.

50 Bingley, *North Wales [...] 1798 and 1801*, II: 345–61.

51 Idem, *Tour round North Wales [...] 1798*, I: iv; idem, *North Wales [...] 1798 and 1801*, I: vi; Jones, *Botanists and Guides*, 84.

52 Moir, *Discovery of Britain*, xv.

53 Susan Major, 'The Million Go Forth: Early Railway Excursions 1840–1860: The Excursion Agent as Social Entrepreneur', www.academia.edu/529584 (accessed 21 November 2012).

54 Ruddy–Pamplin Papers (private archive), Thomas Ruddy, 'Notes on the Life of William Pamplin', 6 ff., draft notes for an article 'William Pamplin: Obituary', September 1899 (newspaper unidentified); William Pamplin, 'Welsh Botany', *Phytologist* new series 2 (1857–58): 312–15.

55 Dewi Jones, '"Nature-Formed Botanists": Notes on Some Nineteenth-Century Botanical Guides of Snowdonia', *Archives of Natural History* 29, part 1 (2002): 31–50 (39–40).

56 Edwin Lees, 'Sketches of a Botanical Ramble in Wales', *Phytologist* 4 (1851): 116–24 (118, 122).

57 See, e.g., idem, 'Observations on the Popularity of Natural History', *The Naturalist* 3 (1838): 115–23, 291–301; Edwin Lees, *The Botanical Looker-Out among the Wild Flowers of England and Wales, etc.*, 2nd ed. (London: Hamilton, Adams, 1851). See Mary M. Jones, *The Lookers-Out of Worcestershire* (Trowbridge & Esher: Worcestershire Naturalists Club, 1980).

58 Lees, *Botanical Looker-Out*, 461–62; Bingley, *North Wales [...] 1798 and 1801*, I: 208–13.

59 David E. Allen, *The Victorian Fern Craze: A History of Pteridomania* (London: Hutchinson, 1969).

60 Lees, *Botanical Looker-Out*, 461.

61 William Bennett, 'Notes on the rarer Ferns observed in a fortnight's Pedestrian Tour in North Wales: with several new Localities for *Asplenium lanceolatum*', *Phytologist* 3 (1847–50): 709–15 (710, 712). See http://herbariaunited.org/wiki/William_Bennett (accessed 7 October 2014).

62 Newman, 'Welch Habitat for *Asplenium germanicum*': 974–75; idem, *A History of British Ferns*, 3rd ed. (London: John Van Voorst, 1854), 77, 268.

63 Allen, *Victorian Fern Craze*, 54–55; idem, 'Changing Attitudes to Nature Conservation: The Botanical Perspective', *Biological Journal of the Linnean Society* 32 (1987): 203–12.

64 Bangor University Archive, Pamplin Papers 25, VI: W. Borrer to W. Pamplin, 11 June 1855; NLW 7507C, Correspondence of William Pamplin, W. Borrer to W. Pamplin, 22 April 1857. The *Lloydia serotina*, the 'Snowdon Lily', named after Lhuyd has been renamed *Gagea serotina*; see Clive A. Stace, *New Flora of the British Isles*, 3rd ed. (Cambridge: Cambridge University Press, 2010), 856.

65 NLW 7509C (1854), 2/x: 3 August 1854.

66 Thomas Pennant, *The Literary Life of the Late Thomas Pennant, Esq., by Himself* (London, 1793), 22; see Jones, *Botanists and Guides*, 48–55.

67 Bingley, *North Wales [...] 1798 and 1801*, II: 363–420.

68 Idem, *Tour round North Wales [...] 1798*, I: x–xii.

69 Ibid. I: 212–13.

70 Bingley, *North Wales [...] 1798 and 1801*, I: 253–54; Jones, *Botanists and Guides*, 80–81.

71 Augustin Ley, 'The Botany of the Honddu and Grwyne Valleys', *Woolhope Club* (1883–85): 343–51 (344).

72 Chater, *Flora of Cardiganshire*, 14–15.

73 Augustin Ley, 'Notes on some of the cliff plants of Wales', *Woolhope Club* (1886–89): 73–86.

74 See Allen, *Books and Naturalists*, 334; Richard Middleton, 'The Royal Horticultural Society's 1864 Botanical Competition', *Archives of Natural History* 41, no. 1 (2014): 25–44; Patrick Armstrong, *The English Parson Naturalist: A Companionship Between Science and Religion* (Leominster: Gracewing, 2000), 54; Ley, 'Notes', 74. The role of private collections is discussed in Samuel J. M. M. Alberti, 'Placing Nature: Natural History Collections and their Owners in Nineteenth-Century Provincial England', *British Journal for the History of Science* 35, no. 3 (2002): 291–311 (293–97).
75 Carter, 'Brecknock', 168, 179.
76 Riddelsdell, 'Lightfoot's Visit to Wales', 290–307.
77 Watson, *Topographical Botany*, 599–606.
78 Alexander Irvine, 'An Epitome of botanical tours in Wales from the earliest period', *Phytologist* new series, 1 (1855–6): 212–19, 264–69.
79 Johnson, *Itinerary of a Botanist*.
80 Louise Miskell, *Intelligent Town: An Urban History of Swansea, 1780–1855* (Cardiff: University of Wales Press, 2012), 167–69, 173–75; Allen, *Naturalist in Britain*, 67.
81 Aikin, *Journal of a tour*, passim; see Arthur H. Dodd, *The Industrial Revolution in North Wales* (Cardiff: University of Wales Press, 1971); John Davies, *A History of Wales* (Middlesex: Allen Lane The Penguin Press, 1993): 319–508; Geraint H. Jenkins, *The Foundations of Modern Wales 1642–1780* (Oxford: Oxford University Press, 1993): 259–426.
82 Dawson Turner and Lewis W. Dillwyn, *The Botanist's Guide through England and Wales* (London, 1805), 31.
83 Allen, *Botanists*, 35–36; Evans, *Tour through part of North Wales*, vi–vii; Turner and Dillwyn, *Botanist's Guide*, 32.
84 Edwin Lees, 'Notice of Plants gathered in the vicinity of Aberystwyth, Cardiganshire', *Phytologist* I (1841–44): 38–40; Carter, 'Cardiganshire', 82.
85 Edwin Lees, 'On the Botanical Features of the Great Orme's Head: with Notices of some Plants observed in other parts of North Wales during the Summer of 1849', *Phytologist* 3 (1847–50): 869–81 (881); see Alfred W. Bennett, 'Notes on the more interesting Flowering Plants gathered in North Wales, in September, 1849', *Phytologist* 3 (1847–50): 771–74; Edwin Lees, 'Notices of the Flowering time and Localities of some Plants observed during an Excursion through a portion of South Devon, in June, 1851', *Phytologist* 4 (1850–3): 530–41 (541).
86 Babington, *Memorials*, 145, 159–62, 318; idem, 'On the Botany of South Pembrokeshire', *Journal of Botany* 1 (1863): 258–70 (259).
87 Bangor University Archive, Pamplin Papers, 8, VII: C. C. Babington to W. Pamplin, 7 Aug[ust] [18]57; Jones, *Botanists and Guides*, 147–50.
88 Alexander Irvine, 'Preface', *Phytologist* new series 6 (1862–3): iii–vi (v); Middleton, 'The Royal Horticultural Society's 1864 Botanical Competition', 27–28.
89 Alexander G. More, 'Plants gathered in the Counties of Pembroke and Glamorgan', *Journal of Botany* 22 (1884): 43–46 (43). Babington, 'Pembrokeshire Plants', *Journal of Botany* 24 (1886): 22–23, believed that agricultural improvements were responsible for the demise of Galingale at this locality.
90 Anon., 'Rhayader and Cwm Elan', *Woolhope Club* (1881–2): 40–49 (41–42).
91 H. Cecil Moore, 'A Visit to the Works of the Proposed Birmingham Water Supply from the Elan Valley in Wales', *Woolhope Club* (1896): 153–80 (153–63); anon., 'Visit to the Works of the Birmingham Water supply from the Elan Valley', *Woolhope Club* (1898–99): 150–57.

92 Thomas Barclay, *The Future Water Supply of Birmingham*, 3rd ed. (Birmingham: Cornish Brothers, 1898), 29–37.

93 George A. Birkenhead, 'The Elan Valley, Radnorshire: Notes on its Natural History and Physical Features', *Reports and Transactions: Cardiff Naturalists Society* 34 (1901–2): 48–52 (49, 51).

94 Mabey, *Gilbert White*, 107.

95 The writer is indebted to Mr Arthur O. Chater for his invaluable comments on an earlier draft of this paper; also to the Revd Mrs Wendy Carey for access to the Ruddy–Pamplin papers; Mrs Elen Wyn Simpson, Assistant Archivist, Bangor University Archive; Mrs Jenny Britnell; and Dr C. Stephen Briggs. Also, the staff of the RCAHMW and NLW and helpful suggestions from anonymous referees.

SHORT BIBLIOGRAPHY OF THOMAS PENNANT'S *TOURS* IN SCOTLAND AND WALES

This bibliography is intended to give the reader an overview of the rather complicated publishing history of Pennant's Scottish and Welsh *Tours*, and may help to clarify certain references or points of detail in the preceding chapters. It makes no claims to completeness: as Hutton and Leask's chapter on the Glenriddell *Tours* shows, printed volumes could be bound and repaginated by their owners, and bibliographical anomalies abound. What this selection does show, however, is the constantly evolving nature of these works, which accrete information, illustrations and appendices with each new edition; it is also testimony to the sheer industry of Pennant, his informants and his publishers in the years 1771–84.

1. *Tour in Scotland* in 1769

(NB 'A Map of Scotland, the Hebrides, and Part of England adapted to Mr Pennant's Tour', engraved by J. Bayly and 'published by Benjamin White, 1st May, 1777', is pasted into many copies of both Pennant's 1769 and 1772 *Tours*, even those published before 1777, suggesting that it was sold as a separate item and pasted in by purchasers.)

a) 1st edition: *A tour in Scotland. MDCCLXIX* (Chester: printed by John Monk, MDCCLXXI [1771]). Octavo, viii + 316 pp, 18 plates, 7 appendices + Itinerary, 316 pp.

b) 2nd edition: *A tour in Scotland. MDCCLXIX* (London: printed for B. White, at Horace's Head, in Fleet-Street, MDCCLXXII [1772]. Octavo, viii + 331 pp. Plates/appendices as in 1st edition.

c) *Supplement to the Tour in Scotland; MDCCLXIX* (Chester: printed by John Monk, MDCCLXXII [1772]). Octavo, 18 pp. This consists of errata, corrections and notes to the Chester edition of the previous year.

d) 3rd edition: *A tour in Scotland; MDCCLXIX* (Warrington: printed by W. Eyres, MDCCLXXIV [1774]. Quarto edition, xiv + 388 pp, 21 plates, 8 appendices (including Itinerary). 'A completely new set of plates and appendices.'

e) *The additions to the quarto edition of the Tour in Scotland, MDCCLXIX. And the new appendix. Reprinted for the accomodation of the purchasers of the first and second editions* (London: printed for B. White, at Horace's Head, Fleet-Street, MDCCLXXIV [1774]. Octavo, 180 pp. With 21 plates, the 7 new appendices and engraved title-page published in the Warrington 3rd edition in quarto.

f) 4th edition: *A tour in Scotland; MDCCLXIX* (London: printed for Benj. White, MDCCLXXVI [1776]) Quarto, 400 pp, 40 plates which combine those from 1st and 3rd editions, appendices as in 3rd edition (although some copies also seem to combine appendices from 1st and 3rd edition)

g) 5th edition: *A tour in Scotland; MDCCLXIX* (London: printed for Benj. White, MDCCXC [1790]) Quarto, 400 pp. Plates as in 4th edition, appendices as in 3rd edition, although again with variations in some copies.

Recent publications: A facsimile reprint edition of the 3rd edition (1774) was published by Melven Press with an introduction by Barry Knight (Perth: Melven Press, 1979); Birlinn Press have reprinted the 1st edition of 1771 with an introduction by Brian D. Osborne (Edinburgh: Birlinn Ltd, 2000).

2. *Tour in Scotland and Voyage to the Hebrides 1772* (Parts I and II)

a) 1st edition (Part I): *A tour in Scotland, and voyage to the Hebrides*; MDCCLXXII [1772] (Chester: printed by John Monk, MDCCLXXIV [1774] Quarto, vii + 439 pp, 44 plates, no appendices but includes Itinerary. (Other copies of this edition consulted have only viii + 379 pp).

b) [? Pirated Dublin edition?] *A tour in Scotland, MDCCLXIX. By Thomas Pennant, Esq*; (Dublin: printed for A. Leathley, [no. 63] in Dame-Street, MDCCLXXV [1775]. 2 vols. Octavo. These seem to be cheap octavo editions without plates. Vol. I (DRY.592, 388 pp) is titled *A Tour in Scotland MDCCLXIX, The Fourth Edition* and contains VIII Appendices, last of which is 1769 Itinerary. Vol. II is entitled *A Tour in Scotland, and Voyage to the Hebrides. MDCCLXXII, The Fourth Edition, Volume II*. This is Part 1 of the 1772 *Tour* (i.e., the *Voyage to the Hebrides*). No appendices, but includes Itinerary of the 1772 *Tour* Part I.

c) 2nd edition (Part I): *A tour in Scotland, and voyage to the Hebrides; MDCCLXXII.* Part I (London: printed for Benj. White, MDCCLXXVI [1776].) Quarto, vii + 439 pp, 44 plates, no appendices but includes Itinerary of 1772 tour.

d) 1st edition (Part II): *A tour in Scotland. MDCCLXXII.* Part II (London: printed for Benj. White, 1776). Quarto, iv + 482 pp, 47 plates and 20 appendices, including Itinerary. This is followed by pp. 1–34 *Additions to the Tour in Scotland* 1769* (including two plates of 'Pictish Houses') paginated continuously with *Additions to the Voyage to the Hebrides 1772* and *Two Omissions, to be referred to their proper place* 'Tour 1769 Scarborough' and 'Voy. Hebrides. Bagpipes'. The index of plates for *A tour in Scotland. MDCCLXXII. Part II* makes reference to 2 Plates of 'Pictish Houses' in the 'Additions' and their pagination, suggesting that the 'Additions' were worked off at the same time as the cover matter for 1776 edition, and were always intended to be bound with the volume.

*NB these are different from the *Additions to 1769* as described at 1e. above.

e) *A tour in Scotland, and voyage to the Hebrides; MDCCLXXII.* 2 vols (London: printed for Benj. White, MDCCXC. [1790]). Quarto. (First complete edition of the 1772 Tour?) Vol. I has vii + 440 pp, and 44 plates. The only appendix is Pennant's Itinerary. Vol. II has iv + 489 pp, and 45 plates, and 21 appendices, including the Itinerary. These are the same as those in the 1st edition (Part II) of 1776, but without the 2 extra plates of 'Pictish Houses' appearing in the *Additions to the Tour in Scotland, MDCCLXIX.* Appendices are slightly different from 1776 edition of 1772 *Tour* Part II.

An edition of the *Tour in Scotland and Voyage to the Hebrides, 1772,* based on Benjamin White's 1776 2nd edition of Part I, and 1st edition of Part II, was published with an introduction by Charles W. Withers, edited by Andrew Simmons (Edinburgh: Birlinn, 1998).

3. *Tour in Wales*

a) 1st edition: *A tour in Wales. MDCCLXX.* (London: printed by Henry Hughes, MDCCLXXVIII [1778]). Quarto, vii + 525 pp, 27 plates.

NB: Pennant's tour in Wales actually took place in 1773. MDCLXX was corrected by hand to MDCLXXIII on a number of copies, but does appear printed as MDCCLXXX in other copies, such as Glasgow's Hunterian K.5.9. and the copy on ECCO.

b) 1st edition: *A Journey to Snowdon* (London: printed by Henry Hughes, 1781), iii + 487 pp, + index and Additions and Corrections to Vol. I of *Tour in Wales*, paginated separately. Often bound with:

c) 1st edition: *Continuation of the Journey* (London: printed by Henry Hughes, 1783)

d) *A tour in Wales. MDCCLXXIII.* (Dublin: printed for Messrs. Sleater, Potts, Moncrieffe, Walker, Exshaw, Flin, Burnet, Jenkin, White and Beatty, MDCCLXXIX [1779].) 469 pp [Vol. 1 only. Probable pirated edition].

e) 2nd edition: *A tour in Wales.* Quarto, 2 vols (London: Printed for Benjamin White, at Horace's Head in Fleet Street, MDCCLXXXIV [1784]). Vol. I, viii + 547 pp. 26 plates; Vol. II, 551 pp, 26 plates, and 17 appendices. These two volumes include all three journeys, and both volumes contain Additions and Corrections, paginated separately.

f) *Tours in Wales by Thomas Pennant Esq.* With notes. 3 vols (London: Printed for Wilkie and Robinson; J. Nunn; White and Cochrane; Longman, Hurst, Rees and Orme; Vernor, Hood, and Sharpe; Cadell and Davies; J. Harding; J. Richardson; J. Booth; J. Mawman; and J. Johnson and Co., 1810). Octavo. The editorial notes are by the author's son, David Pennant. Vol. I, 415 pp; Vol. II, 415 pp; Vol. III, 484 pp. Includes the illustrations by Moses Griffith.

g) *Hynafiaethau Cymreig: teithiau yn Nghymru, sef cyfieithiad o'r "Tours in Wales" gan Thomas Pennant; at yr hyn yr ychwanegwyd cofrestr o bum' llwyth breninol Cymru, a phymtheg llwyth Gwynedd, a'u disgynyddion, fel yr ymddangosasant yn "Pennant's History of Whiteford and Holywell." Ynghyda chyfieithiad o'r nodiadau a'r rhagymadrodd, yn yr argraffiad Seisoneg diweddaf gan John Rhys. Hefyd, nodiadau, hanes bywyd yr awdwr, a rhagarweiniad i hanes y llwythau gan W. Trevor Parkins* (Caernarfon: H. Humphreys, 1883) 635 pp. A translation into Welsh by the Celtic scholar John Rhŷs of the 1810 edition (f, above), with supplementary material on Welsh genealogy added from Pennant's *History of the parishes of Whiteford, and Holywell* (1796).

h) Extra-Illustrated Copy of *A Tour In Wales*. Pennant's own bespoke 8-vol. edition of the 1784 *Tour*, with a wealth of marginal watercolours by Moses Griffith and others, held at NLW and available to view digitally on: https://www.llgc.org.uk/collections/digital-gallery/pictures/journeytosnowdon/

Recent publications: A facsimile edition of *A Tour in Wales*, based on Benjamin White's 2-vol. edition of 1784 (e, above), was published by Bridge Books, with an introduction by R. Paul Evans (Wrexham, 1991); Cambridge University Press have also reprinted this tour (Cambridge, 2014). A limited-edition volume was published by the Gregynog Press (Newtown, 2006): *Pennant and his Welsh Landscapes: Selected Readings from A Tour in Wales (1778–1784)*, edited and introduced by Gwyn Walters, with woodcuts by Rigby Graham.

INDEX

www.ingramcontent.com/pod-product-compliance
Lightning Source LLC
Chambersburg PA
CBHW022350280326
41935CB00007B/142